THE SCIENCE OF CHOCOLATE

RSC Paperbacks

THE SCIENCE OF CHOCOLATE

STEPHEN T. BECKETT

Nestlé Product Technology Centre,
Haxby Road, York YO91 1XY, UK

RS•C

ISBN 0-85404-600-3

A catalogue record for this book is available from the British Library

Reprinted 2006, twice

Published by The Royal Society of Chemistry,
Thomas Graham House, Science Park, Milton Road,
Cambridge CB4 0WF, UK

For further information see our web site at www.rsc.org

Typeset by Paston Prepress Ltd, Beccles, Suffolk, NR34 9QG
Printed in Great Britain by Henry Ling Limited, Dorchester, Dorset, DT1 1HD

Preface

In 1988 I wrote a paper for the School Science Review, in which I described some of the science involved in chocolate making and followed this by two experiments that could be tried in the classroom. As a result of this I received letters from both pupils and teachers requesting more information or new experiments to try. Subsequently I was contacted by Chris Butlin, who was then developing a food option for the Salters' Physics A level course. This resulted in some of the science of chocolate being included in this option. The numerous talks given by my colleagues and myself to junior schools, societies and universities also convinced me that there was a genuine interest in this topic and that people were not just coming for the free samples.

When, therefore, the RSC asked me if I would write a full book on this topic, aimed at schools and universities, I agreed to do so, without realising the amount of work involved. As I graduated in physics, the book is naturally biased in this direction, although I have tried to include a substantial amount of chemistry and even some mathematics in the project work. Several of the chemical terms used in industry are different from those taught in schools. I have attempted to use the current terminology and have included a glossary in the hope that it will be useful if someone is not familiar with the term in the text. This glossary also explains some of the industry's own technical names.

This book should be especially useful for someone studying food science at university or who is about to join the confectionery industry. Although a scientific background is required to understand the more difficult sections, such as fat chemistry or the Maillard reaction, most of the rest of the book should be readable by 16 to 18 year olds. Here I have attempted to show how concepts such as latent heat, relative humidity *etc.* play an important part in the making of something as apparently simple as chocolate. I hope that this in fact might prove to be a 'painless' way of learning about them.

Several sections are relatively simple and can be adapted by teachers of GCSE science or even younger pupils. This is especially true of the projects described in Chapter 10. These are meant to be just basic ideas that can be adapted according to age. All use apparatus or ingredients that should be easy to make or obtain. The appropriate safety precautions must of course be taken, especially for the ones involving glass, heat or chemicals.

Finally I would like to thank my wife Dorothy for her help in reading the draft and indexing the book and our son Richard for his help with the diagrams, together with John Birkett, Richard Crisp, Philippe Gonus, Karen Jones, Huma Lateef, John North, Helmut Traitler and Bronek Wedzicha for correcting the script, producing illustrations or testing the projects to ensure that they worked. I am also grateful to Blackwell Science, Loders Croklaan and Palsgaard Industri A/S for their permission to reproduce diagrams and tables and to Nestlé for approval to publish this book. In particular, Figures 1.2, 2.3, 3.5, 3.6, 3.10, 3.13, 3.14, 4.4, 4.9, 4.10, 4.11, 5.2, 5.3, 5.8, 5.10, 5.13, 6.8, 7.1 and 7.4 are all reproduced from 'Industrial Chocolate Manufacture and Use', with the kind permission of Blackwell Science.

Contents

Glossary

Acetic acid: Common name for ethanoic acid.

Chocolate crumb: A dehydrated mixture of milk, sugar and cocoa liquor, used as an ingredient for some types of milk chocolate.

Cocoa butter: Fat pressed out from the centre (nib, cotyledon) of cocoa beans.

Cocoa butter equivalent: A fat which can be mixed with cocoa butter in any proportion without upsetting the way it crystallises.

Cocoa liquor: Cocoa nib which is finely ground. Like chocolate it is solid at room temperature but liquid above 35 °C.

Cocoa mass: Another name for cocoa liquor.

Cocoa nib: Cocoa beans with the shell removed.

Conche: A machine which mixes the chocolate ingredients to make it into a liquid and to remove some of the unwanted flavours.

Ethanoic acid: Also known as acetic acid.

Ethanedioic acid: Also known as oxalic acid.

Enrober: A machine which coats sweet centres, by pouring chocolate over them.

Lauric fat: Fat rich in dodecanoic (lauric) acid (C12:0). It is a major component of fats from coconut and palm kernel.

Outer: Box containing a number of bars or tablets.

Oxalic acid: Also known as ethanedioic acid.

Phosphoglyceride: A fat containing phosphoric acid (or other phosphorus-containing acids) in appropriate ester form such as glycerophospholipids.

Phospholipid: Commonly used name for phosphoglyceride.

Plastic viscosity: A measurement related to the viscosity of a liquid when it is moving relatively quickly.

Polymorphism: The ability of a substance to crystallise in several different forms, with different melting points.

Tempering: A process for ensuring that the fat in the chocolate sets in the correct crystalline form.

Triacylglycerol: A class of fats made up of glycerol esterified to three fatty acids.

Triglyceride: Commonly used name for triacylglycerol.

White chocolate: Chocolate made from cocoa butter, sugar and milk powder.

Yield value: A measurement relating to the energy required to start a liquid flowing, *i.e.* its viscosity when it is moving very slowly.

Chapter 1

The History of Chocolate

Chocolate is almost unique as a food in that it is solid at normal room temperatures yet melts easily within the mouth. This is because the fat in it, which is called cocoa butter, is mainly solid at temperatures below 25 °C when it holds all the solid sugar and cocoa particles together. The fat is, however, almost entirely liquid at body temperature, enabling the particles to flow past one another, so the chocolate becomes a smooth liquid when it is heated in the mouth. Chocolate also has a sweet taste that is attractive to most people.

Strangely chocolate began as a rather astringent, fatty and unpleasant tasting drink and the fact that it was developed at all is one of the mysteries of history.

CHOCOLATE AS A DRINK

The first known cocoa plantations were established by the Maya in the lowlands of south Yucatan about 600 AD. Cocoa trees were being grown by the Aztecs of Mexico and the Incas of Peru when the Europeans discovered central America . The beans were highly prized and used as money as well as to produce a drink known as chocolatl. The beans were roasted in earthenware pots and crushed between stones, sometimes using decorated heated tables and mill stones, similar to the one illustrated in Figure 1.1. They could then be kneaded into cakes, which could be mixed with cold water to make a drink. Vanilla, spices or honey were often added and the drink whipped to make it frothy.[1] The Aztec Emperor Montezeuma was said to have drunk 50 jars of this beverage per day.

Christopher Columbus bought back some cocoa beans to Europe as a curiosity, but it was only after the Spaniards conquered Mexico that Don Cortez introduced the drink to Spain in the 1520s. Here sugar was added to overcome some of the bitter, astringent flavours, but the drink remained virtually unknown in the rest of Europe for almost a hundred

Figure 1.1 *Ancient decorated mill stone with a hand grinder from the Yucatan*

years, coming to Italy in 1606 and France in 1657. It was very expensive and, being a drink for the aristocracy, its spread was often connected to connections between powerful families. For example the Spanish princess, Anna of Austria, introduced it to her husband King Louis XIII of France and the French court in about 1615. Here Cardinal Richelieu enjoyed it both as a drink and to aid his digestion. Its flavour was not liked by everyone and one Pope in fact declared that it could be drunk during a fast, because its taste was so bad.

The first chocolate drinking houses were established in London in 1657 and it was mentioned in Pepys's *Diary* of 1664 where he wrote that 'jocolatte' was 'very good'. In 1727 milk was being added to the drink. This invention is generally attributed to Nicholas Sanders.[2] During the 18th century White's Chocolate House became the fashionable place for young Londoners, while politicians of the day went to the Cocoa Tree Chocolate House. These were much less rowdy than the taverns of the period. It remained however, very much a drink for the wealthy.

One problem with the chocolate drink was that it was very fatty. Over half of the cocoa bean is made up of cocoa butter. This will melt in hot water making the cocoa particles hard to disperse as well as looking unpleasant because of fat coming to the surface. The Dutch, however, found a way of improving the drink by removing part of this fat. In 1828 Van Houten developed the cocoa press. This was quite remarkable, as his entire factory was manually operated at the time. The cocoa bean cotyledons (known as cocoa nibs) were pressed to produce a hard 'cake' with about half the fat removed. This was milled into a powder, which could be used to produce a much less fatty drink. In order to make this powder disperse better in the hot water or milk, the Dutch treated the cocoa beans during the roasting process with an alkali liquid. This has subsequently become known as the Dutching process. By changing the type of alkalising agent, it also became possible to adjust the colour of the cocoa powder.

EATING CHOCOLATE

Having used the presses to remove some of the cocoa butter, the cocoa powder producers were left trying to find a market for this fat. This was solved by confectioners finding that eating chocolate could be produced by adding it to a milled mixture of sugar and cocoa nibs. (The ingredients used to make dark chocolate are shown in Figure 1.2.) If only the sugar and cocoa nibs were milled and mixed together they would produce a hard crumbly material. Adding the extra fat enabled all the solid particles to be coated with fat and thus form the hard uniform bar that we know today, which will melt smoothly in the mouth.

Almost twenty years after the invention of the press, in 1847, the first British factory to produce a plain eating chocolate was established in Bristol in the UK by Joseph Fry.

Unlike Van Houten, Fry used the recently developed steam engines to power his factory. It is interesting to note that many of the early chocolate companies, including Cadbury, Rowntree and Hershey (in the USA) were founded by Quakers or people of similar religious beliefs. This may have been because their pacifist and teetotal beliefs prevented them working in many industries. The chocolate industry was, however, regarded as being beneficial to people. Both Cadbury and Rowntree moved to the outside of their cities at the end of the 1890s, where they built 'garden' villages for some of their workers. Fry remained in the middle of Bristol and did not expand as quickly as the other two companies. It eventually became part of Cadbury.

Figure 1.2 *A picture of the unmilled ingredients used to make dark chocolate:* A, *sugar;* B, *cocoa butter;* C, *cocoa nibs* (Beckett[3])

With the development of eating chocolate the demand for cocoa greatly increased. Initially much of the cocoa came from the Americas. The first cocoa plantation in Bahia in Brazil was established in 1746. Even before this, the Spaniards had taken cocoa trees to Fernando Po (Biyogo), off the coast of Africa, and this soon became an important growing area. In 1879 a West African blacksmith took some plants home to the Gold Coast (now Ghana). The British Governor realised its potential and encouraged the planting of trees, with the result that Ghana has become a major source of quality cocoa. Other European powers also encouraged the growing of cocoa in their tropical colonies, *e.g.* France in the Ivory Coast (Côte D'Ivoire), which is now the world's largest producer of cocoa.

The chocolate made by Fry was initially a plain block and it was only in 1876 that the first milk chocolate was made by Daniel Peter in Switzerland. Chocolate can not contain much moisture, because water reacts with the sugar and turns melted chocolate into a paste rather than a smoothly flowing liquid (see Project 5 in Chapter 10). As little as 2% of moisture can give a product a poor shelf-life as well as an inferior texture. This meant that Daniel Peter had to find some way of drying the plentiful supply of liquid milk that he found in his own country. He was helped in this by the recent development of a condensed milk formula by Henri Nestlé. This meant that he had much less water to evaporate. Also he was able to remove the remaining moisture using relatively cheap water-powered machines. In most countries milk chocolate products are now much more popular than plain chocolate ones.

In order for the chocolate to feel smooth on the tongue, when it melts in the mouth, the solid non-fat particles must be smaller than 30 microns (μm; 1000 microns = 1 mm). The chocolates made by Fry and Peter were ground using granite rollers, which could not mill as fine as this, so they still had a gritty texture. Groups of particles also joined together to form agglomerates, and some of the fat did not coat the particles very well owing to poor mixing. All these effects gave the chocolate a poor texture. In addition the chocolate tended to taste bitter because of the presence of some acidic chemicals (see Chapter 4).

In 1880 Rodolphe Lindt, in his factory in Berne in Switzerland, invented a machine which produced a smoother, better tasting chocolate. This was known as a conche, because its shape was similar to that of the shell with that name (Figure 1.3). It consisted of a granite trough, with a roller, normally constructed of the same material, which pushed the warm liquid chocolate backwards and forwards for several days. This broke up the agglomerates and some of the larger particles and coated them all with fat. At the same time moisture and some acidic chemicals were evaporated into the air, producing a smoother, less astringent tasting chocolate. A schematic diagram of the chocolate making processes is shown in Figure 1.4.

Figure 1.3 *Picture of chocolate being processed in a long conche*

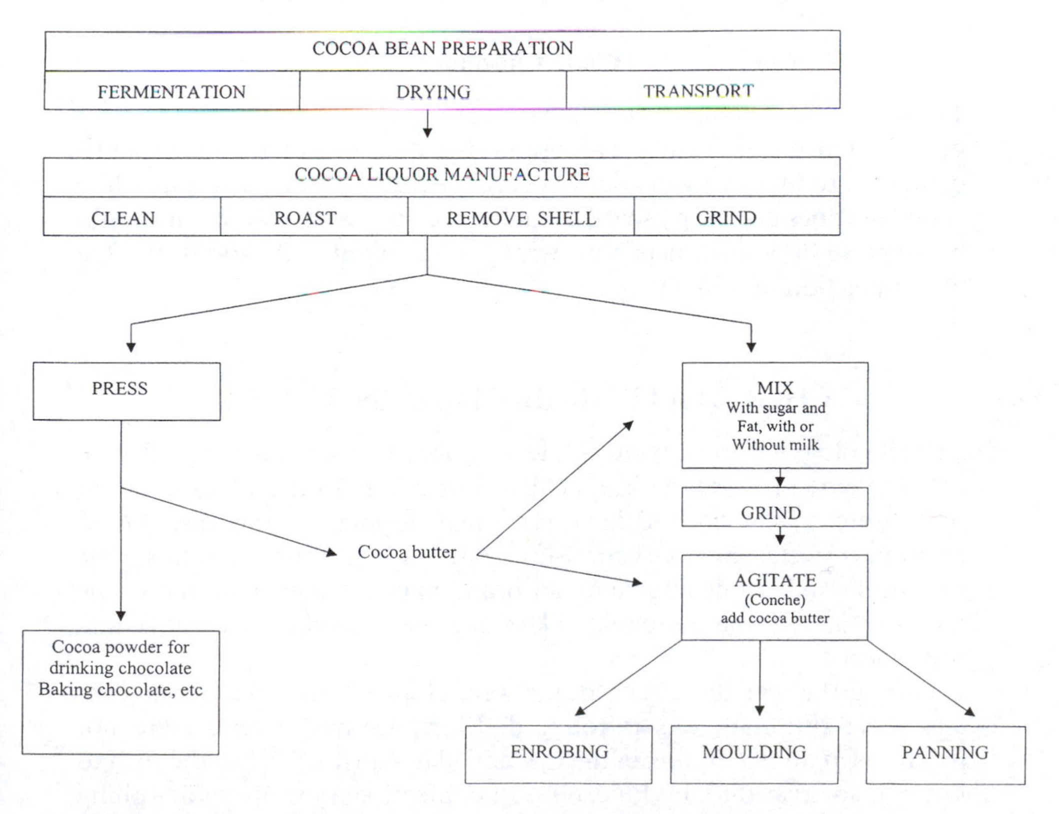

Figure 1.4 *Schematic diagram of the chocolate manufacturing process*

Chocolate Crumb

In the early part of the 20th century the milk used to make chocolate had poor keeping qualities. This caused problems for the chocolate industry, whose major sales were at Christmas, a time of the year when there was a very limited supply of fresh milk. In the UK and some other countries this led to the development of an intermediate ingredient called chocolate crumb.

The cocoa nibs contain substances known as antioxidants. These restrict the breaking up of the fats, which would normally make the milk fat turn sour. In addition sugar was known to extend the shelf-life of foods (it is used in jams *etc.*). The chocolate manufacturers therefore added sugar and cocoa to the milk and dried them together. This produced chocolate crumb, which had a shelf-life of at least a year. Milk produced during the spring peak could then be used to make chocolate the following Christmas. The drying process, however, introduced some cooked flavours into the chocolate and it is for this reason that many UK chocolates taste different from some continental European ones, which are made from milk powder.

White Chocolate

The first white chocolate was made in 1930. It was made from sugar, milk powder and cocoa butter. The preserving qualities of the cocoa antioxidants are mainly in the dark cocoa material. This means that white chocolate does not keep as well as milk chocolate, and also that it should be kept in a non-transparent wrapper, as light will speed up the decomposition of milk fat.

CHOCOLATE MARKETING IN THE UK

As technology improved, chocolate was used to coat other ingredients, or to be part of a product rather than just a bar. In the 1930s many of these were developed and have remained popular to this day. Good examples of this are KitKat, Mars Bar and Smarties. At this time products also became known under brand names rather than that of the manufacturer. Some companies, like Cadbury, tended to give both equal prominence.

During the war few cocoa beans were shipped from the plantations and strict rationing was introduced. Many leading brands were not produced at all. Rationing in the UK ended in April 1949, but the rush to buy was so great that, by June, 60% of confectionery shops had nothing left to sell. Rationing was reimposed until February 1953.

Consumption rose very quickly, but over the last ten years has been more constant with an average of about 8 kg/person per annum of chocolate confectionery being eaten in many Western European Countries (this does not include chocolate biscuits). Switzerland has the highest consumption at about 9.5 kg/person per annum. This makes the confectionery industry a very important one. The combined sales of sugar and chocolate confectionery in the UK are more than those of tea, newspapers and bread put together.[4]

CHOCOLATE IS GOOD FOR YOU

Antioxidants in food are known to protect the body against chemicals called free radicals that damage cells. Cocoa is a known source of antioxidants and in 1999 doctors from the National Institute of Public Health and the Environment in Bilthoven in the Netherlands examined chocolate for its catechins content. These are from the family of flavonoids, which are among the most powerful antioxidants. They found that dark chocolate contained 53.5 mg per 100 g, which is four times that in tea. Drinking a cup of tea a day is said to reduce the chance of a heart attack.[5]

Certain fats are said to cause an increase in cholesterol. Because of its triglyceride structure (see Chapter 6) cocoa butter has been found to be neutral in this respect. Studies by Prof. Kris-Etherton in the USA, where high risk individuals were fed large amounts of chocolate, showed that their cholesterol level remained unaffected.[6]

REFERENCES

1. R. Whymper, 'Cocoa and Chocolate. Their Chemistry and Manufacture', Churchill, London, 1912.
2. L.R. Cook (revised by E.H. Meursing), 'Chocolate Production and Use', Harcourt Brace, New York, 1984.
3. S.T. Beckett (ed.), 'Industrial Chocolate Manufacture and Use', 3rd Edition, Blackwell, Oxford, UK, 1999.
4. Nestlé, 'Sweet Facts '98', Nestlé Rowntree, York, UK, 1999.
5. Anon 'Antioxidants in Chocolate', *Manufacturing Confectioner*, (Sept.) 1999, 8.
6. P.M. Kris-Etherton, 'Dietary Fatty Acids and Cardiovascular Disease Risk', 43rd Technology Conference, BCCCA, London, 1996.

Chapter 2

Chocolate Ingredients

COCOA BEANS

Cocoa Trees

To be called chocolate, a product must contain cocoa. The cocoa or cacao tree (*Theobroma cacao*, L.) originated in South and Central America, but is now grown commercially in suitable environments between 20° north and 20° south. These areas have a high average temperature (⩾27 °C) throughout the year and a constant high humidity, arising from a plentiful rainfall (1500–25 000 mm). The soil should be deep and rich and well drained and normally be less than 700 m above sea level, as strong winds will damage the crop.

The trees are relatively small, 12–15 m in height, and grow naturally in the lower level of the evergreen rainforest. In commercial plantations they are often sheltered by inter-cropping trees such as coconut and banana. The leaves are evergreen and are up to about 300 mm in length. Trees start bearing pods after 2 to 3 years, but it is 6 or 7 years before they give a full yield.

There are three types of cocoa. Criollo has beans with white cotyledons and a mild flavour. The trees are, however, relatively low yielding. Most cocoa is Forastero, which is more vigorous and often grown in West Africa in smallholdings (a family's cultivated land that is smaller than a farm). The third form, Trinitario, is usually thought to be a hybrid of the other two types.

The trees are attacked by many pests and diseases Some of the most serious are:

- capsids (insects that feed on sap, causing damage to plant tissue)
- black pod disease (fungi which attack mainly the pod, making them rot)
- witches' broom disease (fungal attack causing growths or brooms to develop from leaf buds, but also affects flowers and pods)

- cocoa pod borer moth (larva bores into the pods and affects the development of the beans)

Commercial Cocoa-Producing Countries

There are three major cocoa growing regions: West Africa, South-East Asia and South America. These are shown in Figure 2.1. The cocoa supplies from individual countries has changed dramatically over the last twenty years due to economic changes as well as pests and diseases.

Bahia in Brazil was a major cocoa growing area producing over 400 000 tonnes in the mid 1980s. It now produces less than half of this, largely due to destruction by witches' broom disease. Instead of being a major exporter of beans, almost all are now processed and used within Brazil.

Over 40% of the world's crop is now grown in the Ivory Coast (Côte d'Ivoire). Production has increased dramatically over the last 30 years and a large proportion of European chocolate it made from this source of beans.

Ghana is the second highest producer with around 14% of the world's crop. It also has a good reputation for producing quality beans. Nigeria produces some cocoa, but the establishment of the oil and other industries resulted in alternative employment and a reduction in cocoa production.

Malaysia built up a fairly big cocoa production in the 1980s, but this has declined rapidly, partly due to pod borer infestation and also due to the greater profitability of other crops such as oil palm.

Indonesia is currently expanding its cocoa growing and is now producing almost as much as Ghana.

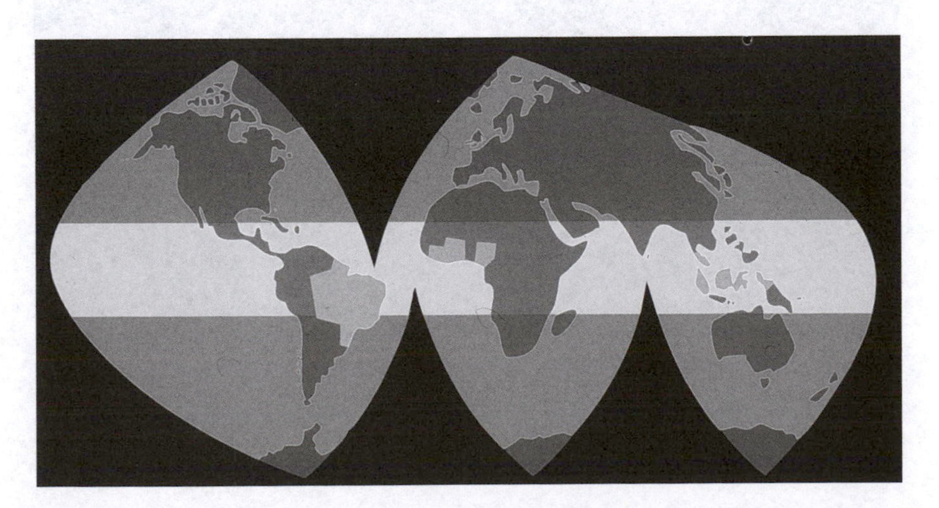

Figure 2.1 *Map showing the major cocoa growing countries of the world*

The flavour of the cocoa depends not only on the cocoa type, *e.g.* whether it is Criollo or Forastero, but also upon the climate and soil conditions *etc*. For some specialist chocolates, normally dark ones, beans are obtained from specific areas. These flavour cocoas, often Criollo, are produced in many smaller growing areas such as Ecuador, the Caribbean Islands and Papua New Guinea.

In addition to the flavour of the beans, the fat contained within it also changes according to the area of production. In general the nearer the equator that the tree is grown the softer is the fat, *i.e.* the easier it is to melt. This means that Malaysian cocoa butter is relatively hard, whereas most Brazilian cocoa butter is much softer. The harder is better for chocolates, which will be sold in summer, whereas the softer is preferable for frozen products, such as choc-ices, where the fat is hardened by the cold conditions. See Chapter 6 and Project 10 in Chapter 10.

Cocoa Pods

Tiny flowers (Figure 2.2), up to 100 000 in number, grow on the branches and trunk of the tree throughout the year. These grow into small green pods called cherelles (Figure 2.3), but take 5–6 months to develop into mature pods (Figure 2.4) between 100 and 350 mm long. Their weight ranges from 200 g to more than 1 kg and they exist in a wide variety of shapes and colours depending upon variety. Each pod contains some 30–45 beans.

Figure 2.2 *Picture of flowers on cocoa trees*

Figure 2.3 *The small pods or cherelles growing on trunk of a cocoa tree* (Fowler[1])

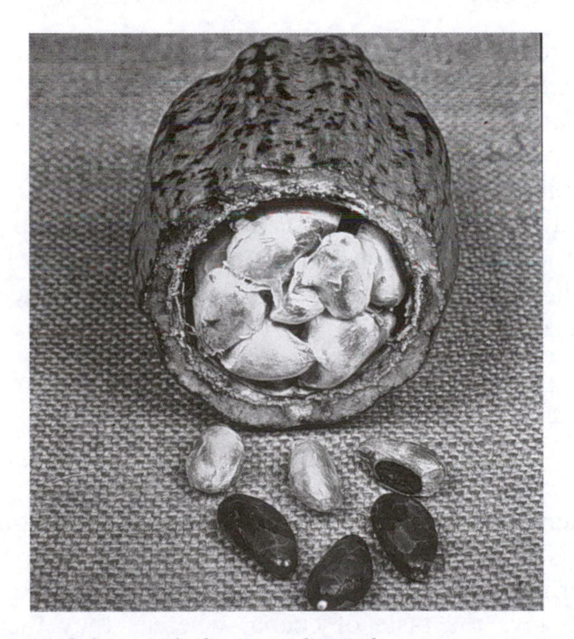

Figure 2.4 *Cocoa pod showing the beans on the inside*

The pods are carefully cut off the tree with a machete (cutlass), where they are within reach. For the higher branches it is necessary to use a special knife attached to a long pole, as is shown in Figure 2.5. Pods are normally harvested every 2–4 weeks over a period of several months, as they do not all ripen at the same time.

Figure 2.5 *Cocoa pods being gathered by a long pole*

The pods are opened with a machete or cracked with a wooden club. The beans are oval in shape and covered in a white pulp (mucilage). The beans are separated from the majority of this pulp by hand.

The beans consist of an outer shell or testa surrounding two cotyledons (called nibs) and a small germ (the embryo plant). The cotyledons store the food for the developing seedling and also its first two leaves. Much of the food is in the form of a fat (cocoa butter) which accounts for over half of the dry weight of the bean. The moisture content of the bean at this stage is about 65%.

Fermentation

Fermentation Procedure

Correct fermentation is essential to produce a good flavour in the final chocolate. It is a process in which the bean is killed, so that it can not be spoiled by germination. In addition certain chemicals are formed, which upon heating give the taste of cocoa, whereas they themselves taste completely different, or may not even taste of anything at all. These are known as flavour precursors as they lead to the flavour, but aren't that flavour themselves. Unfermented beans may be pressed to produce cocoa butter, but the remaining solid cocoa material is not normally used to make chocolate.

A lot of cocoa trees are grown by smallholders and the method of fermentation is traditional, although in some countries there have been

Figure 2.6 *Cocoa beans being fermented under banana leaves*

attempts to modernise it. There are two main types of method: heap and box fermentation.

In West Africa, heap fermentation is widely used. Between 25 and 2500 kg of fresh beans, together with a small amount of the white pulp, are placed in a heap and then covered with banana leaves (Figure 2.6). The process normally lasts from 5 to 6 days, with the actual length being determined by experience. Some farmers turn the beans after 2 or 3 days. The smaller heaps often produce the better flavours.

The larger plantations, particularly in Asia, use the box fermentation technique. The wooden boxes may hold between 1 and 2 tonnes of beans, which are designed with outlet holes or slits, usually in the base (Figure 2.7). These provide ventilation and let the water, which comes out of the beans and pulp, run away. These may be up to a metre deep, but shallower ones (250–500 mm) often give a better flavour due to the improved ventilation. The beans are tipped from one box to another each day to increase aeration and give a more uniform treatment. Usually the fermentation period is similar to that for the heap procedure, although some plantations may take a day or two longer.

Microbial and Chemical Changes

What actually happens during fermentation has been the subject of much research and is still not entirely clear. In addition, as the bean shell remains intact, it is not possible for micro-organisms to react directly with the cotyledons inside, which are the part that is used to make chocolate. So, in a way, this is not a true fermentation process at all.

Figure 2.7 *Cocoa beans being fermented in boxes*

During the fermentation, however, the temperature rises dramatically during the early stages and three days of heat are thought to be sufficient to kill the bean. Following its death, enzymes (a type of catalyst capable of greatly increasing the rate of break down of substances like fat into simpler components, but which comes out of the reaction unchanged itself) are released. These cause the rapid decomposition of the beans' food reserves and form sugars and acids, which are the precursors of the chocolate flavour.

The process is, however, much more complicated than this, as a more usual fermentation process takes place outside the bean. Here there is some of the white pulp, which is very sugar-rich and able to react with the yeasts that are also present to form acids and ethanol, much in the same way as occurs during brewing. This ethanol activates other bacteria, *e.g.* ethanoic (acetic) and lactic acid bacteria, which then convert it into their respective acids. The ethanol and acids are able to pass through the shell into the bean. This change in acidity (pH) hastens the death of the bean.

The different ways of fermenting will also give rise to different flavours. For example in box fermentation the beans are moved every day. This aerates the beans and stimulates those bacteria that require oxygen (*e.g. Acetobacter*) and encourages the production of ethanoic acid. Other reactions, for instance those involving yeasts, are retarded by the presence of oxygen, so less ethanol is formed. This means that cocoas that are box fermented are more likely to taste acidic than the same type of cocoa that has been heap fermented. In order to overcome some of this acidity, some box fermentation processes shorten the fermentation time and reduce the number of turnings.

Many other important reactions also occur. The proteins and peptides react with polyphenols to give the brown colour associated with cocoa, whilst other flavour precursors are formed by reactions between sucrose and proteins. Of particular importance is the formation of amino acids. Proteins are composed of a series of amino acids joined sequentially by peptide bonds. Amino acids can be represented as:

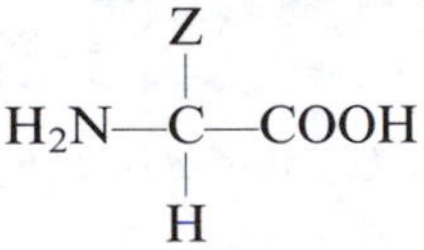

$$
\begin{array}{c}
Z \\
| \\
H_2N\text{—}C\text{—}COOH \\
| \\
H
\end{array}
$$

$$Z = H, CH_3, C_2H_5, CH_2OH \textit{ etc.}$$

(In practice it is ionic with either the H_2N protonated or the COOH dissociated depending upon the pH of the surroundings.)

Many of the proteins break up during the fermentation into these acids. 20 forms exist, including valine and glycine, which are very important in chocolate flavour formation.

Further details of chemical changes during fermentation are given by Fowler[1] and Dimick and Hoskin.[2]

Drying

Following fermentation the beans must be dried before they can be transported to the chocolate making factories. Failure to do this will result in moulds growing on the beans. These give the chocolate a strong nasty flavour and so can not be used. Beans must also not be over-dried. Those with a moisture content of less than 6% become very brittle, which makes subsequent handling and processing much more difficult.

Where the weather permits, the beans are usually sun dried. They are spread out during the day in layers about 100 mm thick on mat, trays or terraces. They are raked at intervals and heaped up and protected at night or when it rains. In Central and South America a roof on wheels is used to cover up the beans, which are laid on the floor. In Ghana split bamboo mats are placed on low wooden tables. The mats can be rolled up when it rains. Here it normally takes about a week for the beans to dry to the required 7–8% moisture level, which is too low for moulds to grow. In other areas beans are dried on moveable tables that can be put under cover when necessary (Figure 2.8). A major problem of sun drying

Figure 2.8 *Cocoa beans being dried on moveable tables*

is the risk of contamination from the surroundings and from farm and wild animals wandering amongst the beans. This means that precautions must be taken in handling them when they reach the chocolate making factory (see Chapter 3).

In other countries, particularly in Asia, the weather may be too wet and artificial drying is required. Sometimes wooden fires are lit in a chamber below the drying area and the hot gas led through a flue beneath the drying platform then out through a vertical chimney. A major problem here is that of smoke leaking from the flue. This, like mould, gives the beans an unpleasant harsh flavour and prevents them from being used for chocolate making. Forced-air dryers are better, as are efficient heat exchangers that stop smoky contaminants reaching the beans. If the drying is too quick, the beans will taste very acidic and it is better to dry them at lower temperatures or intermittently over a longer period.

Storage and Transport

The beans must be stored so that they do not pick up water, as they will become mouldy once their moisture level rises above 8%. Traditionally they are stored in 60–65 kg jute sacks (Figure 2.9). These are strong, stackable and allow the moisture to pass through. They are also biodegradable. As chocolate is a very delicate flavour, the beans must also be stored well away from other goods such as spices, which might result in off-flavours in the chocolate.

The beans are often transported in the holds of ships. At the point of loading the temperature will be about 30 °C, but very soon the temperature in the North Atlantic will be nearer freezing point. If the beans have

Figure 2.9 *Cocoa in jute sacks ready for loading onto a ship*

a moisture content of 8% their equilibrium relative humidity is about 75%. In other words, if the relative humidity is below 75% the beans will dry out but they will pick up moisture at higher humidities. The moisture level must not exceed 8%, so beans must never be stored at higher humidities. In the ships, however, this is difficult. The humidity is already high when the sacks are loaded and the drop in temperature causes a rapid rise in the relative humidity to 100% (the dew point). Moisture will condense on the ship's structure and sometimes get into the sacks making the beans mouldy. The sacks should therefore never be in contact with the cold surfaces of the ship's sides and absorbent mats should be placed on top of them. The ship's hold should also be ventilated to remove the moist air.

SUGAR AND SUGAR SUBSTITUTES

Traditionally chocolate has been made containing about 50% sugar, mostly in the form of sucrose, but with some lactose from the milk components in milk chocolate. Diabetics are unable to eat much sugar, so other recipes were developed to incorporate fructose (a different form of sugar, also found in honey) or non-sugar bulk sweeteners such as sorbitol. More recently there has been a requirement for lower calorie or 'tooth friendly' chocolates, so other sugar substitutes have been developed.[3]

Sugar and Its Production

Sucrose (also known as saccharose) is produced from both sugar beet and sugar cane. Both give the same natural crystalline disaccharide material. It is called disaccharide because it is composed of two single sugars (monosaccharides) chemically linked together. These sugars, called glucose and fructose, are in equal parts and can be separated by acidic treatment or by using an enzyme called invertase. The resulting mixture of the two sugars is called invert sugar (equation 2.1).

$$\underset{\text{Sucrose}}{\overset{\textit{Sugar}}{C_{12}H_{22}O_{11}}} + H_2O \rightarrow \underset{\substack{\text{Glucose + Fructose} \\ \text{(isomers)}}}{\overset{\textit{Invert sugar}}{2C_6H_{12}O_6}} \tag{2.1}$$

Lactose is also a disaccharide and is made up of a combination of glucose and galactose. Many of the sugar substitutes, like sorbitol, are sugar alcohols.

Sugar beet contains about 14–17% sucrose. The beet is cleaned and sliced and the sugar, together with some mineral and organic impurities are washed out by warm water. Slaked lime is added to precipitate out these impurities and carbon dioxide is then bubbled through the solution. This precipitates out the excess slaked lime as calcium carbonate. The solids are removed by filtration to give a 15% sugar solution, which is then evaporated to 65–70%. Vacuum evaporation and centrifuges are then used to purify and crystallise the sucrose. It is not possible to recover all the sugar in a single processing stage and white sugar requires three or four different steps.

Sugar cane has a sucrose content of 11–17%. The raw juice is squeezed from the crushed stalks, often using roller mills. The remaining material can be used to make paper, cardboard or hardboard *etc*. This juice contains more invert sugar than was the case with beet sugar. This makes it more difficult to crystallise and a gentler treatment is required to get rid of the impurities otherwise an undesirable brown colour would form. Alternative treatments involve lower temperatures or the use of sulfur dioxide. Hydrocyclones and bow-shaped sieves are used to clarify the liquid and crystallisation procedures are similar to those used for beet sugar.

Crystalline and Amorphous Sugar

Crystal sugar is extremely pure, normally being more than 99.9% and rarely less than 99.7%. It can be purchased with different crystal sizes, which are approximately as follows:

coarse sugar	1.0–2.5 mm grain size
medium fine sugar	0.6–1.0 mm grain size
fine sugar	0.1–0.6 mm grain size
icing sugar	0.005–0.1 mm grain size

Most chocolate manufacturers use medium fine sugar, although some ask for defined particle size spectra, as this may aid the flow of the final chocolate (see Chapter 5).

All these sugars are in a crystalline form. Sucrose crystals can in fact be grown to be several centimetres long and can take many crystalline forms. All are, however, birefringent. This means that if the crystals are placed in a polarising microscope, with the polaroid filters set so that no light is transmitted, they bend the light and are seen as bright images (Figure 2.10).

Sugar can also exist as a glass, *i.e.* a non-crystalline though solid structure. A good example is a clear boiled sugar sweet, such as a Glacier mint. This happens when sucrose solutions are dried too quickly and the individual molecules do not have time to form the crystalline structure when the water is removed. One way to manufacture amorphous sugar is to freeze dry a sucrose solution. Amorphous sugar is not birefringent as it does not possess a structure such that it can bend the light in a microscope. There are other ways of determining amorphous sugar in sucrose systems (see Project 1 in Chapter 10).

Amorphous sugar is important in chocolate making as it can affect both the flavour and the flow properties of liquid chocolate. Its surface is very reactive and can easily absorb any flavours that are nearby. It is also formed from crystalline sucrose at high temperatures. These may occur when sugar is milled. If there is no other material around the sugar may take up a metallic note. (This can be demonstrated by finely grinding

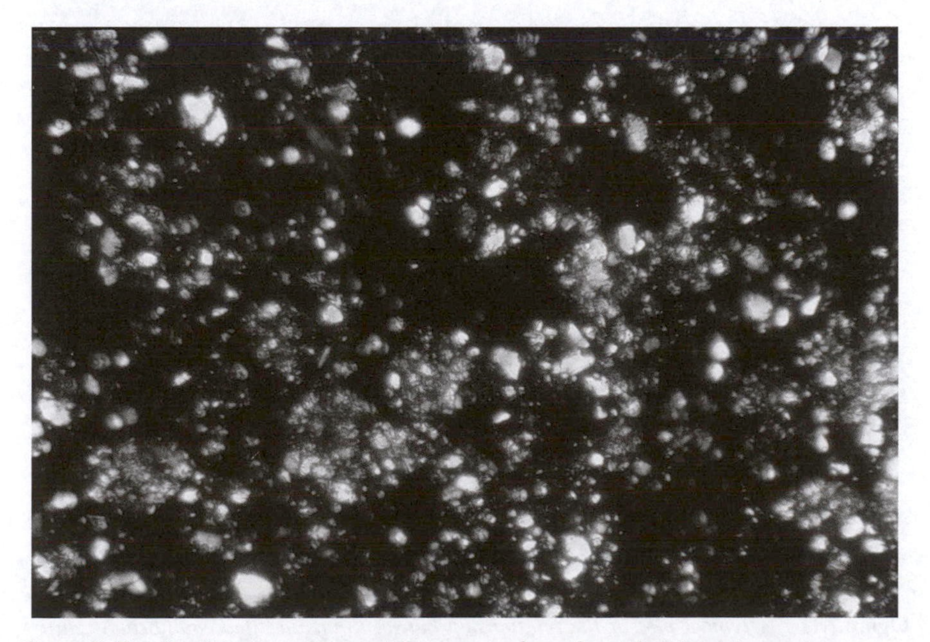

Figure 2.10 *Birefringent sugar particles as seen through a polarising microscope*

sugar in a food mixer and then dissolving the sugar in water: it will taste metallic compared with a solution made from the original material.) On the other hand, if it is milled together with cocoa, some of the volatile cocoa flavours are absorbed by the amorphous sugar rather than escape into the atmosphere as they would otherwise do. This will then produce a more intense flavour chocolate. Care must be taken when milling sugar, especially by itself, because of the high risk of an explosion.

The amorphous state is an unstable one, and in the presence of water it will turn into crystalline material. Once the change has taken place the moisture is expelled, as crystalline sucrose is essentially anhydrous. About half the mass of chocolate is sucrose, so the particles within it are very close together. The moisture on the surface makes them stick together. This builds up a sugar skeleton, which holds together even if the fat melts. This is the basis of a method used to create a chocolate suitable for sale in hot climates. If the chocolate has not yet been solidified, the stickiness on the surface of some of the sugar greatly increases the viscosity of the liquid chocolate.

Crystalline sugar can also absorb moisture, depending upon its surrounding conditions. The storage conditions that should be used can be determined by means of sorption isotherms. Figure 2.11 illustrates the curve for sugar at 20 °C. As was noted earlier, the equilibrium relative humidity is the relative humidity at which water is neither taken in nor given out. This means that between 10% and 60% the sugar will

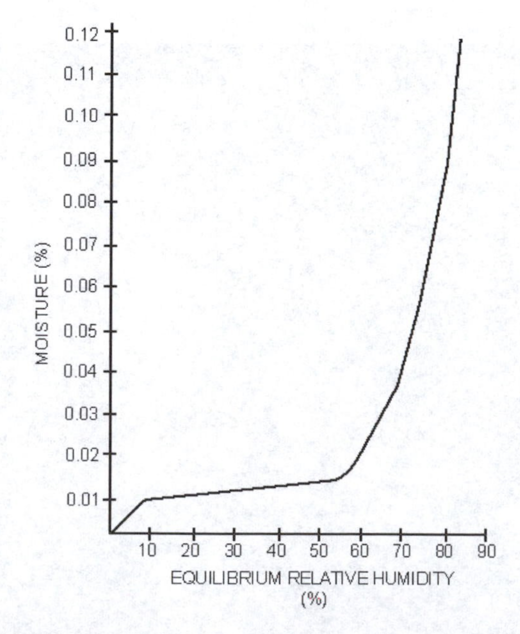

Figure 2.11 *A typical plot of the relationship between moisture and equilibrium relative humidity for crystal sugar*

maintain a moisture of between 0.01% and 0.02%. At higher humidities, the moisture content increases dramatically. Damp sugar can be microbiologically contaminated. In addition it will stick together and form lumps, even if the humidity is reduced again.

In the chocolate industry, sugar is stored in large silos containing many tonnes. Great care must be taken with the storage conditions as otherwise the silo will block up and fail to empty. Very often the air inside them is de-humidified.

Lactose

Like sucrose, lactose is a disaccharide, this time consisting of monosaccharides called glucose and galactose. It is part of cow's milk (see next section) and is therefore found in all milk chocolates. It is sometimes used as crystalline lactose to replace part of the sucrose. As it is much less sweet than normal sugar, it reduces the overall sweetness of the chocolate. The crystalline form is a monohydrate, *i.e.* it contains one molecule of water, which it does not lose even if the temperature is raised to 100 °C. There are in fact two crystalline forms α- and β-lactose. The α-form is produced by most conventional processes and is slightly less sweet and less soluble than the other form. Both are non hygroscopic, *i.e.* they do not readily take up water. The amorphous form is, however, very hygroscopic.

When lactose is added in the form of spray-dried milk powder it is normally in an amorphous form. This can be demonstrated by Project 1 in Chapter 10. This glassy state is able to hold in some of the milk fat and so make it unavailable to help the chocolate flow (Chapter 5).

At elevated temperatures lactose is able to take part in what is called the Maillard or browning reaction. This is the type of reaction that occurs to produce cooked flavours when toasting food. It will be described in more detail in Chapter 3.

Some people are lactose intolerant, *i.e.* their body reacts against this type of sugar, and they must limit the amount of milk products that they eat.

Glucose and Fructose

Although these monosaccharides (single sugars) combine together to make the disaccharide sucrose, they are not normally used to make chocolate. Glucose, also called dextrose, crystallises as a monohydrate and is very difficult to dry completely. It normally contains some water and also rapidly absorbs it from the surrounding air (*i.e.* it is very hygroscopic). This moisture makes the liquid, melted chocolate very thick, because it tends to stick the sugar particles together.

Fructose is also a very hygroscopic. It is found naturally in fruits and honey. It is sometimes found in special chocolates for diabetics as, unlike sucrose, it does not raise the blood sugars when eaten. It does, however, need special processing conditions, especially with regard to temperature and humidity.

Sugar Alcohols

Sugar alcohols (polyols) are used to replace sucrose in chocolate, when it is required to make a lower-calorie or a sugar-free product. Sucrose is normally regarded as containing 4.0 kcal g^{-1} (17 kJ g^{-1}). Although the different sugar alcohols probably have different calorific values, in Europe, for legislative purposes, they are regarded as having 2.4 kcal g^{-1} (10 kJ g^{-1}). Like fructose, they are suitable for diabetics, but unlike fructose they are suitable to make non-cariogenic chocolate, in other words chocolates that are not damaging to the teeth. One of them, xylitol, which is found naturally in many mushrooms and fruits, can not be fermented by most of the bacteria in the mouth and in some countries is regarded as beneficial to the teeth.

Other common sugar alcohols include sorbitol, mannitol, isomalt and lactitol. Some require to be processed into chocolate at relatively low temperatures, to prevent them forming gritty lumps. All tend to have a laxative effect. The EU Scientific Committee on Foods recognised this to be important, but stated that a consumption of 20 g per day was unlikely to have a harmful effect.

There is also a big difference in sweetness between the different sugar alcohols. The relative value with respect to sucrose is given in Table 2.1. Intense sweeteners, such as aspartame, may be required with some polyols such as sorbitol, to off-set their lower sweetness.

When some substances dissolve in water, energy is required to enable this to take place. This occurs for sugar (see Project 1 in Chapter 10) and

Table 2.1 *Relative degree of sweetness of different sugars and sugar alcohols* (from Krüger[3])

Sugar	*Relative sweetness*
Sucrose	1.0
Fructose	1.1
Glucose	0.6
Xylitol	1.0
Maltitol	0.65
Sorbitol	0.6
Mannitol	0.5
Isomalt	0.45

Table 2.2 *Relative cooling effect of bulk sweeteners* (from Krüger[3])

Sugar	*Relative cooling effect*
Sucrose	1.0
Polydextrose	−2.0
Lactitol (anhydrous)	1.4
Isomalt	2.1
Maltitol	2.5
Fructose	2.6
Lactitol (monohydrate)	3.0
Sorbitol	4.4
Xylitol	6.7

results in a cooling of the water to enable the molecules to separate and dissolve. This takes place to a much greater degree with the sugar alcohols. Table 2.2 shows the relative degrees of cooling, which is particularly noticeable for xylitol. This property is often regarded as being undesirable, as chocolate is not expected to have a cooling effect.

Polydextrose

Polydextrose is a polysaccharide (*i.e.* made up of several sugars), which is found in many low-calorie chocolate bars. It has a legislative calorific value in Europe of 1 kcal g^{-1}, that is less than half that of the sugar alcohols, and it has much less of a laxative effect. It is made up from the single sugar glucose with small amounts of the sugar alcohol sorbitol. Because it is amorphous it reacts with water to give out heat (just like amorphous sucrose, see Project 1 Chapter 10). This means that it gives a warming sensation when dissolved in the saliva in the mouth. It also dries out the mouth, sometimes making the chocolate harder to swallow.

MILK AND OTHER DAIRY COMPONENTS

In most countries of the world much more milk chocolate is bought and eaten than both dark and white put together. It tends to be softer than dark chocolate and has a creamier taste and texture.

The majority of cow's milk is water but, as was mentioned earlier, moisture destroys the flow properties of liquid chocolate so only the anhydrous components can be used. Typically these form about 13.5% of the liquid milk and their composition is illustrated in Figure 2.12. The largest component, at just under 5%, is lactose, the disaccharide sugar, described in the previous section. There is almost the same amount of milk fat and about 3.5% of protein. Minerals account for

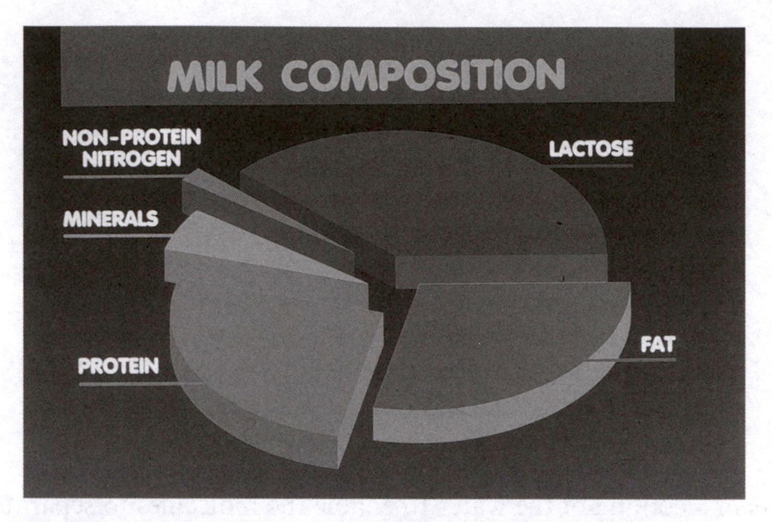

Figure 2.12 *The relative composition of dried milk*

about 0.7%. Calcium in particular is regarded as being very beneficial to health.

Milk Fat

This is the second largest component in dehydrated milk and is vital in giving milk chocolate its distinctive texture and flavour release. It also changes its snap and can inhibit the formation of the white powdery surface on chocolate, which is composed of large fat crystals and is known as fat 'bloom' (see Chapter 6).

The more fat that is present in a liquid chocolate the easier it will flow, both when making sweets and in the mouth. It is also relatively expensive, so the manufacture needs to make the best use of the fat present.

Butter fat is almost entirely liquid at room temperature, so there is a limit to the amount that can be added to chocolate for it still to remain hard. To make matters worse there is a phenomenon, called fat eutectics, which means that when two fats are added together the resultant mixture is often softer than would be expected. This is described in more detail in Chapter 6. This softness does, however, reduce the waxiness in the mouth if harder fats than cocoa butter are present. Fractionate milk fats are also available. These are made by separating out the higher or the lower melting point fractions. Certain ones are said to produce a harder chocolate than the normal milk fat, whereas others are thought to give better bloom prevention.

The fat is 98% triacylglycerols (*triglycerides*), *i.e.* three acids combined with a glycerol molecule. The remaining significant component is

Table 2.3 *Fatty acid composition of milk fat* (Haylock *et al.*[4])

Fatty acid	*Weight (%)*
C4:0* Butyric	4.1
C6:0 Caproic	2.4
C8:0 Caprylic	1.4
C10:0 Capric	2.9
C10:1 Decenoic	0.3
C12:0 Lauric	3.5
C14:0 Myristic	11.4
C16:0 Palmitic	23.2
C18.0 Stearic	12.4
C18.1 Oleic	25.2
C18:2 Linoleic	2.6
C18:3 Linolenic	0.9
Others	10.0

* The first number is the number of carbon atoms in the molecule and the second is the number of double bonds.

phosphoglyceride (*phospholipid*) (mainly lecithin) as well as diacylglycerols (*diglycerides*) (two acids combined with glycerol) and sterols. A typical fatty acid composition in milk is given in Table 2.3.

Milk fat, however, has a limited shelf-life as it can be oxidised or attacked by enzymes (lipolysis). The enzymes accelerate the break-up of the acids into shorter chain free acids, which have a rancid off-flavour and make the chocolate unacceptable. When this type of reaction occurs with cocoa butter, however, the acids formed are largely tasteless, so the chocolate remains acceptable.

The initial result of oxidation is the formation of peroxides (containing O_2 groups). These have no taste themselves but decompose to produce unpleasant off-flavours. A measurement of the amount of peroxide present is used to detect the early stages of deterioration. In order to keep the milk fat for longer periods, contact with oxygen must be minimised. Sometimes the air in the packaging is replaced by nitrogen and oxygen barrier packaging is used. Chilled storage is preferred and the presence of copper and iron must be avoided as these act as catalysts for the oxidation process.

Milk Proteins

Not only do these add to the nutritional content of the chocolate, they are also important in determining its flavour, texture and liquid flow properties. A milk chocolate has a creaminess, which depends to a large extent upon the balance between these proteins and the more acidic

Table 2.4 *Properties of milk proteins*

Caseins	*Whey proteins*
Water binding	Soluble over a wide pH range
Heat stable	Denatured by heat
Open-chain structure	Globular
Fat binding	

flavour from the beans. If the protein proportion is reduced the product becomes much less creamy. Also, like lactose, if they are subjected to water and heat, they can take part in the Maillard (browning) reaction, which introduces cooked flavours into the chocolate.

There are two very different types of protein involved, namely caseins and whey proteins. There are four to five times as much caseins as there are whey proteins. Some of their properties are summarised in Table 2.4.

Caseins act as emulsifiers, *i.e.* as interfaces between two different media. In chocolate this is likely to be between the solid and fat components (*cf.* lecithin in Chapter 5). Its actual role is not understood, but calcium caseinate has been shown to produce thinner chocolate, whereas whey proteins make the same chocolate much thicker. The water binding properties of the caseins will also be beneficial to the chocolate flow.

Although helping the flavour of some chocolates, the casein flavour itself is not particularly pleasant and may be undesirable in other confectionery products.

Milk Powders

Milk can be dried to produce a wide range of different powders. Figure 2.13 shows a flow chart of the dairy processes used to make powders for chocolate making.

The most common powders used for chocolate making are skim milk and full cream milk powder. With the former, milk fat is added at the chocolate making stage so that both powders can be used to make chocolates with the same overall milk components content. They will, however, have different flavours, textures and liquid flow properties. This is in part due to different heat treatments during the drying, but also due to the different state of the fat. With skim milk and milk fat, all the fat is free to react with the particles and cocoa butter, whereas many full cream milk powders have fat tightly bound within the individual particles. This means that there is less fat to help the flow or to soften the cocoa butter, when full cream powders are used.

For many years the milk powders were dried on hot rollers. These machines were expensive and hard to maintain in a hygienic condition,

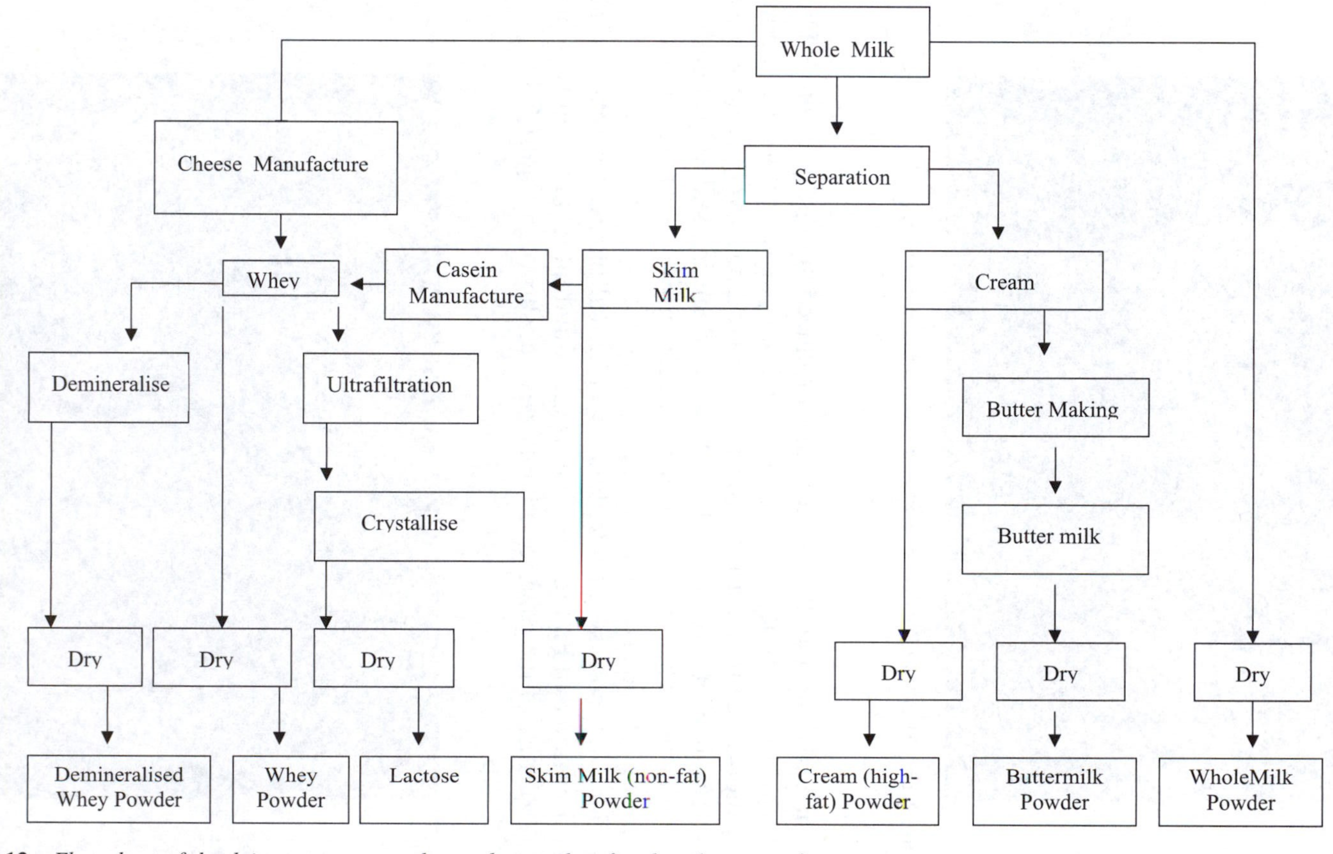

Figure 2.13 *Flow chart of the dairy processes used to make powders for chocolate manufacture*

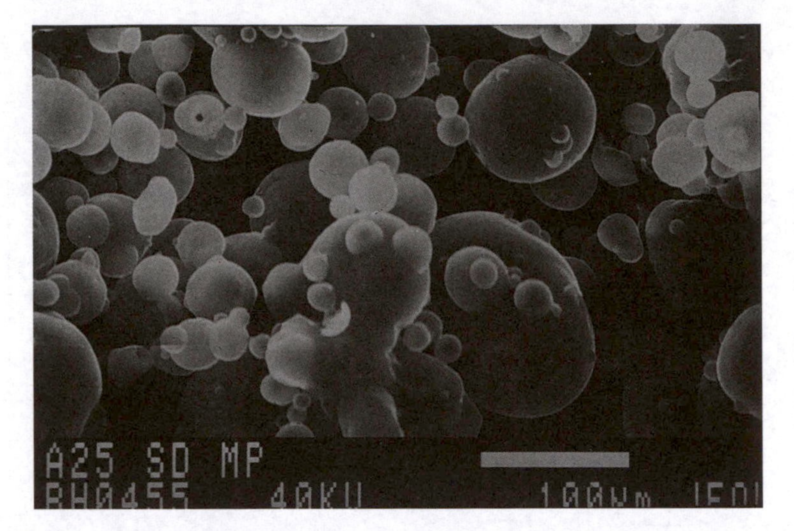

Figure 2.14 *Electron microscope picture of spray-dried skim milk powder*

so currently most milk powders are produced by spray drying. This involves converting the milk, which has been partly pre-dried to about 50% moisture, into a mist ('atomised') so that it has a very large surface area. The droplets are exposed to a flow of hot air in the drying chamber. The air provides heat to evaporate the water and also acts as a carrier for the powder, which is then collected using cyclones or filter systems. This powder often consists of fine, hollow, spherical particles similar to the ones illustrated in Figure 2.14. This is very different from the more plate-like structure of the roller-dried milk powder shown in Figure 2.15.

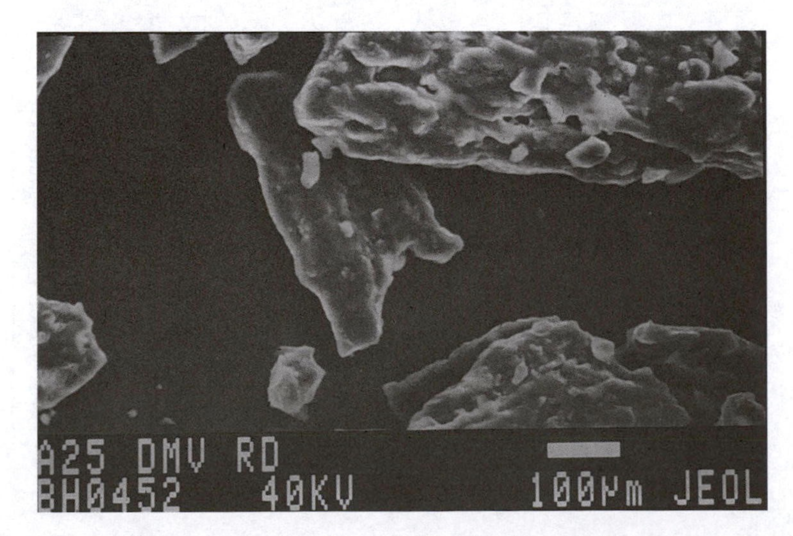

Figure 2.15 *Electron microscope picture of roller-dried skim milk powder*

When the fat in a chocolate is liquid, these particles must pass by one another when the chocolate flows, *e.g.* when it moves around the mouth. The differently shaped particles will flow past each other differently, which may result in a change of viscosity. The spherical particles also contain fat inside them, whereas the rollers press most of it onto the surface. This means that roller-dried full cream milk powder chocolates are softer and flow more easily than spray-dried ones. Modifications can be made to the spray-drying procedure to make the particles more crystalline so as to enable them to release the fat more easily.

The milk may also be heat treated before spraying, to introduce a more cooked flavour. Sometimes extra fat is added to give a high-fat powder, so that the chocolate manufacturer has no need to make subsequent additions during the production stage. Other powders contain sugar and are a form of chocolate crumb.

Whey and Lactose Powders

Both these powders are used in the making of certain types of chocolates and flavoured coatings. They are by-products of cheese and casein manufacture and have the advantage that they are less sweet than sugar and so can replace some of it when a reduction in sweetness is required.

The acidity (pH) of the cheese making greatly affects the mineral content of the whey powder. Normally the more acidic the whey, the higher will be the mineral content. This may give rise to off-flavours in the chocolate and demineralised whey powders may be preferable.

If the protein is removed from the whey, lactose is left. This is then concentrated and crystallised.

CHOCOLATE CRUMB

As was noted in Chapter 1, chocolate crumb exploits the antioxidant properties in cocoa to produce a milk-containing chocolate ingredient with a long shelf-life. It also gives a slightly cooked flavour to the chocolate.

There are many ways of producing crumb, but all require the final product to have a moisture content of between 0.8% and 1.5%. At this level the water activity (equilibrium water activity divided by 100) is so low that micro-organisms can not grow. Care must be taken as to where it is stored, because, as with sugar, it will pick up water under moist surroundings.

Some manufacturers dissolve sugar in fresh or concentrated milk, whereas others add water to a mixture of sugar and milk powder. Multiple effect plate dryers may be used to dry this to 80–90% solids. Milled cocoa beans (cocoa liquor) are then mixed with this material, which is then dried under vacuum in either a batch or a continuous process. At this stage there is heat, moisture, lactose and proteins. These are ideal conditions for the Maillard reaction to occur. This gives the crumb a brown colour and a caramel type flavour. This flavour is quite different from the one which would be obtained if sugar were heated alone. Also subsequent processes have almost no moisture present, so once again it is not possible to get this type of flavour.

Most milk chocolates that are available on the market throughout the world are made from very similar ingredients, yet have very different flavours. Some manufacturers have specific 'house flavours', Cadbury and Hershey perhaps being the most well-known. Very often this is due to specific changes during the crumb making process, where changes in holding times, acidity and temperature can give rise to very different flavours in the final chocolate. For this reason, chocolate crumb making conditions are often kept a closely guarded secret.

REFERENCES

1. M.S. Fowler, 'Cocoa Beans: From Tree to Factory', in S.T. Beckett (ed.), 'Industrial Chocolate Manufacture and Use', 3rd Edition, Blackwell, Oxford, UK, 1999.
2. P.S. Dimick and J.C. Hoskin, 'The Chemistry of Flavour Development in Chocolate', in S.T. Beckett (ed.), 'Industrial Chocolate Manufacture and Use', 3rd Edition, Blackwell, Oxford, UK, 1999.
3. Ch. Krüger, 'Sugar and Bulk Sweeteners', in S.T. Beckett (ed.), 'Industrial Chocolate Manufacture and Use', 3rd Edition, Blackwell, Oxford, UK, 1999.
4. S.J. Haylock and T.M. Dodds, 'Ingredients from Milk', in S.T. Beckett, 'Industrial Chocolate Manufacture and Use', 3rd Edition, Blackwell, Oxford, UK, 1999.

Additional reading

G.A.R. Wood and R.A. Lass, 'Cocoa', 4th Edition, Longman, 1985.

Chapter 3

Cocoa Bean Processing

Traditionally cocoa beans were transported to the country where the chocolate was going to be manufactured, which was normally situated in a temperate climate. Increasingly, however, the cocoa growing countries are processing their own beans to produce cocoa liquor. This has the advantage that the liquor is much easier to transport – the problem of moisture affecting the beans on board ships was noted earlier. In addition, the shell, which is essentially a waste product, is not transported thousands of miles just to be thrown away. The disadvantage is that the manufacturer has less control over the times and temperature of the bean processing, which have an effect on the final chocolate flavour. One solution is to partially process the beans, with the final heat treatment being carried out at the chocolate factory. Wherever the process is carried out, it involves cleaning the beans, removing the shell and some form of roasting.

BEAN CLEANING

As many of the beans are dried on the ground, they often contain sand, stones, iron, plant material *etc*. These must be removed for two reasons. Firstly many of these impurities are very hard and will damage the machinery, which is used to grind the beans. Secondly the organic contaminants will burn during the roasting process and give off gases likely to spoil the cocoa flavour. The cleaning is therefore carried out at the beginning of the chocolate making process.

Normally there are several different procedures, which are combined to remove the various types of waste materials. Iron is removed by magnets, whereas dust can be drawn off by suction. Stones may be a similar size to the beans, but they are of a different density. They can be separated by vibrating them together on a grid, which is set at an angle to the horizontal. Air passes through the grid and blows the beans higher than the stones. As they are nearer the vibrating grid, the stones are moved towards the top, where they fall off into a collecting bag. The air transports the beans towards the lower part of the grid, from where they go to the next stage of processing.

ROASTING AND WINNOWING

The cotyledons (nib) must be roasted, before they can be made into chocolate. This is to change the flavour precursors into the chemicals which actually taste of chocolate. In addition the high temperature, coupled with the remaining moisture in the beans, will kill any microbiological contaminants, such as *salmonellae*, which may be present on the beans from when they were dried on the open ground.

Many manufacturers use a concept known as hazard analysis of critical control points (HACCP). This means that the whole chocolate making procedure is evaluated to determine any possible source of harm to the consumer. As the beans are purchased from bulk suppliers and may be contaminated with harmful bacteria *etc*., it is necessary to treat all the beans as if they were a potential hazard, until the risk has been removed. The roasting procedure will do this, so once this has been carried out the cocoa is absolutely safe. Analysis of the beans at this stage, known as a critical control point, will confirm this.

The main risk is then that the contamination from the untreated beans will be transferred to the rest of the factory. For this reason, the preroasting procedures, such as cleaning, are normally carried out in a separate building. The operators are also obliged to change their clothing before entering the rest of the factory.

The Problem of Bean Size Variation

Cocoa beans come in a variety of sizes, depending upon the country of origin, climatic conditions, the season of the year the pod was picked, and numerous other factors. Traditionally the beans were roasted in small batches of perhaps a few hundred kilograms often in ball shaped roasters (see Figure 3.1). The operator could remove a few beans from the roaster or the attached cooling tray, and from the smell of the beans was able to adjust the temperature and times to ensure that the correct flavour had been developed. Beans could also be segregated so that they were of a similar size within each batch. With modern factories, which need to process several tonnes of beans per hour, this is no longer possible.

The problem caused by having a variety of bean sizes is illustrated in Figure 3.2. Here it can be seen that, when the roasting conditions are set for the average size of bean, the smaller ones are over-roasted, whereas the centre of the larger beans is not roasted enough. This means that in

Figure 3.1 *Ball roaster for cocoa beans*

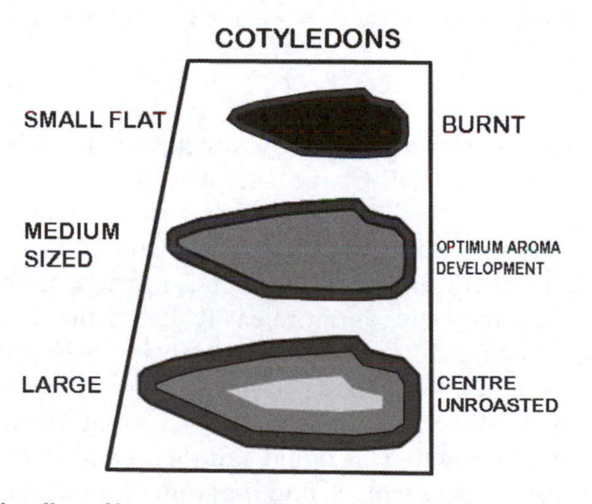

Figure 3.2 *The effect of bean size upon the degree of roasting*

the latter case not all the flavour precursors will have been converted, so the chocolate flavour will be low. In the case of the small bean, further compounds may have been produced, which are not necessarily of the required flavour.

The difficulty with the large beans is illustrated in Figure 3.3. Here the temperature is measured in the roaster itself and at several locations within a bean. Even after a relatively long time (15 minutes) the temperature of the centre of the bean fails to reach that of the outside. The difference in formation of the different flavour compounds can also

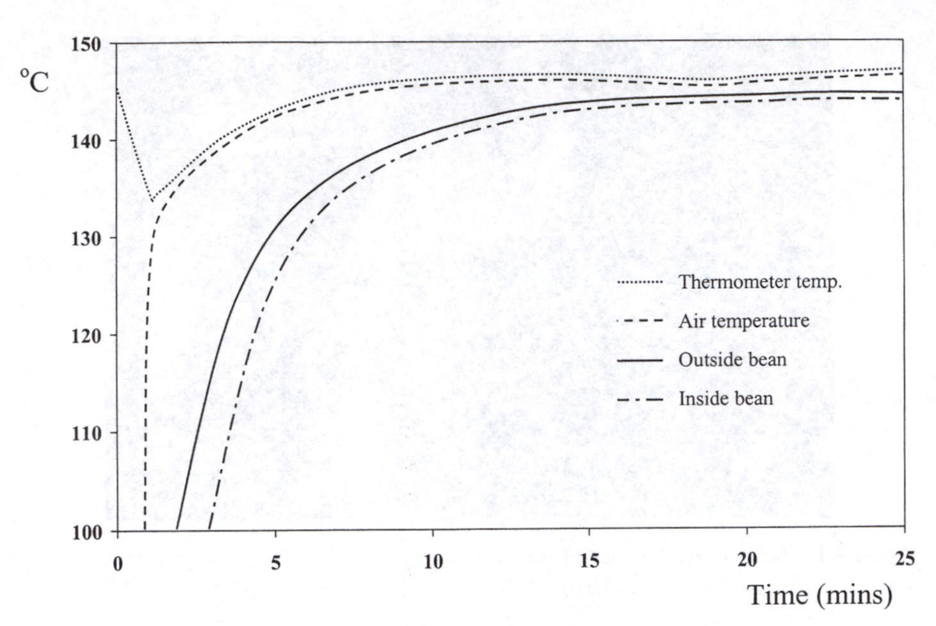

Figure 3.3 *The change of temperature in the roaster and within a bean during the early part of roasting*

be shown by carrying out analysis by systems such as high-pressure liquid chromatography (HPLC, see Chapter 8) on different sizes of bean.

In order to overcome this, two alternative methods have been developed. In the first only the centre of the bean is roasted, so the pieces are much smaller, and the heat can more easily get to the middle. This is known as nib roasting. In the second the nib is finely ground and is turned into cocoa liquor (also known as mass). Here the cocoa butter has been released from the cells within the bean, so that when it becomes warm it turns into a liquid. This liquid can then be heated in a process called liquor roasting. A schematic flow diagram of the different roasting methods is given in Figure 3.4.

Both of these methods require the shell to be removed before the roasting process. This must be done carefully, because the shell often has silica particles attached, often originating from when the beans were dried on the ground. These, together with the shell, which is also very hard, will damage any grinding equipment during the chocolate making. In addition there are legal requirements, which state that only a small amount of shell can be present in anything that is sold as chocolate. The shell is also said to give an inferior flavour to the chocolate. The small amount of shell fat present is different from cocoa butter, and like milk fat (Chapters 2 and 6) has a softening effect on the chocolate.

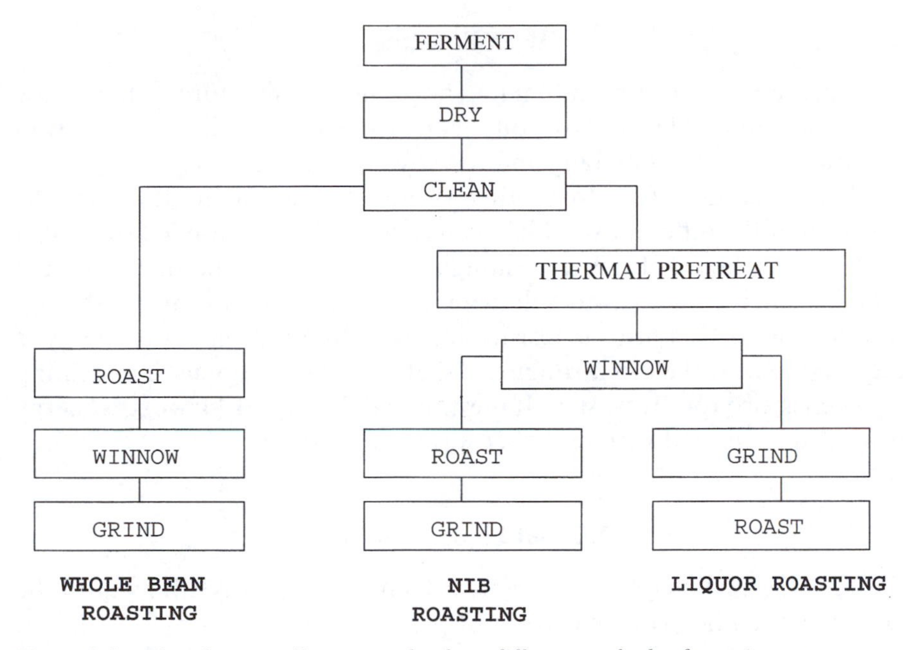

Figure 3.4 *Flow diagram illustrating the three different methods of roasting cocoa*

Winnowing

Winnowing is the process of separating the shell and some of the germ from the rest of the bean. As its name suggests, it relies on the same principles as are used to separate the corn from the chaff during grain harvesting.

It is desirable to keep the central cotyledons (nibs) in as large pieces as possible so that they can separate more easily from the shell. Any small pieces that stay with the shell will be thrown away with it, so economically it is very important to carry out winnowing correctly.

Any broken beans are initially separated, to prevent them being broken further, and go straight to separation process. The remaining beans are broken, often by sending them individually at high speed against impact plates. These, too, then go to the vibrating sieves.

The shell is largely fibrous material and is normally in the form of a flat platelet. The nib, on the other hand, is normally much more spherical and being over half fat is much denser. When the two are vibrated together the lighter shell will come to the top (*cf.* the separation of beans and stones). If air is drawn upwards through this mixture, the lighter shell, with its larger surface area, will rise, whereas the heavier nib falls for further processing. This principle is illustrated in Project 2 in Chapter 10.

Bean Roasting

This process is still used by many chocolate manufacturers. It has the major advantage that the roasting tends to help separate the shell from the nib. This makes breaking and winnowing relatively easy.

There are, however, two other disadvantages in addition to the problem of the different sized beans. When the heat is applied the cocoa butter melts. Some of it is free enough to migrate into the shell, where it remains and is thrown away following winnowing. It is estimated that up to 0.5% of cocoa butter is lost in this way. In addition extra energy is required to heat the nib through the shell. All the energy used in heating the shell is also totally wasted. It is estimated that up to 44% extra energy is needed compared with the other forms of roasting.

Nib and Liquor Roasting

Because the shell is relatively firmly attached to the cotyledons until the bean has been heated, some form of pre-treatment is normally required before winnowing. This normally consists of exposing the beans very rapidly to a source of intense heat provided by saturated steam or infrared radiation. This heats the surface, but the centre remains much cooler and no flavour changing reactions occur. Water inside the bean evaporates and puffs out the shell, making it separate much more easily when the bean breaks during the cracking procedure.

In the case of nib roasting, the machines used are very similar to that for bean roasting. For liquor roasting, however, the nib must be finely milled to turn it into a liquid. This requires very careful moisture control. If the moisture content is too high the cocoa liquor will be a thick paste and not a liquid. Even small amounts of water react strongly with this cellulose–protein–fat system and a moisture content of about 10% would produce an almost solid material which would be very difficult to grind. Very low moisture contents would, on the other hand, produce a chocolate with a poor flavour. During roasting the flavour precursors can react in different ways, depending upon the amount of moisture present. At very low moistures they are unable to produce the desirable reaction compounds.

Roasters

The process may be a batch or continuous one. Instead of the ball-shaped roaster (Figure 3.1) a drum shape is more commonly used (Figure 3.5). These can process as many as 3 tonnes of beans in a single batch. The heat can be applied externally through the walls or by passing hot air through the drum.

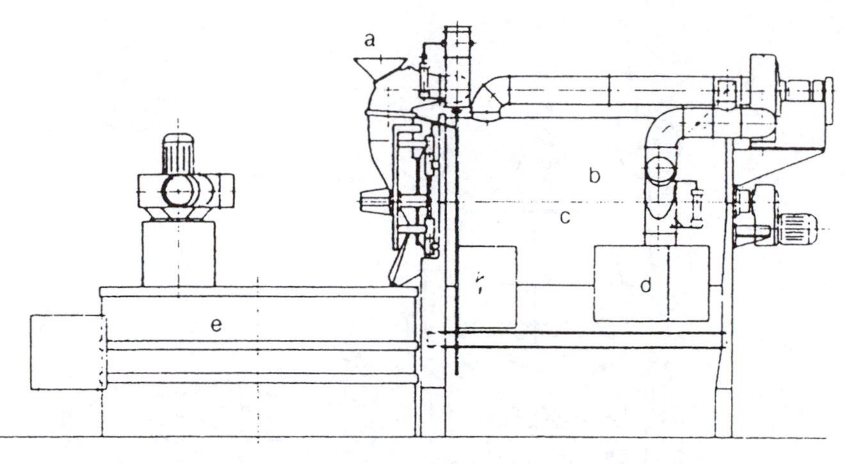

Figure 3.5 *Schematic representation of a batch roasting unit (Barth Ludwigsburg GmbH & Co., Germany).* (a) *Funnel tube;* (b) *reaction drum;* (c) *perforated pipe to carry the solution for alkalising;* (d) *gas or oil heating;* (e) *cooling pan* (Heemskerk[1])

In order to kill the microbiological contaminants both water and heat are required. These roasters are designed to add water or steam so as to increase the effectiveness of the kill. Care must be taken, however, to dry the beans again before the roasting proper takes place, as too much water can remove desirable as well as undesirable flavour components. As mentioned previously, too dry beans are also undesirable. Normally the temperature is raised to its roasting level of between 110 °C and 140 °C when the moisture level has been reduced below 3%. The total roasting procedure normally lasts between 45 minutes and 1 hour. After roasting the product is usually cooled in an external cooler.

Where it is necessary to process large quantities of beans or nibs, a continuous process is often used. A typical continuous roasting system is illustrated in Figure 3.6. Here the beans are fed in batches through the top onto a shelf system. This shelf is made up of a series of slats, through which the hot air blows. After a predetermined time, the slats tilt in turn, starting at the bottom one, until the top one has returned to its position and a new batch has been fed into the top. In this way the beans fall on to the shelf below and through the roaster. The bottom shelves are used to give rapid cooling. Very large volumes of hot air pass through these roasters and care must be taken to ensure that this does not take with it some of the more volatile flavours, together with the moisture that evaporates during the roasting process.

Liquor roasting is carried out in specially designed equipment, which spreads the hot liquid cocoa liquor over the surface of a long hot cylinder with a fast rotating central column. Rotors and paddles, attached to the column, continuously stir the liquid and scrape the surface to stop the

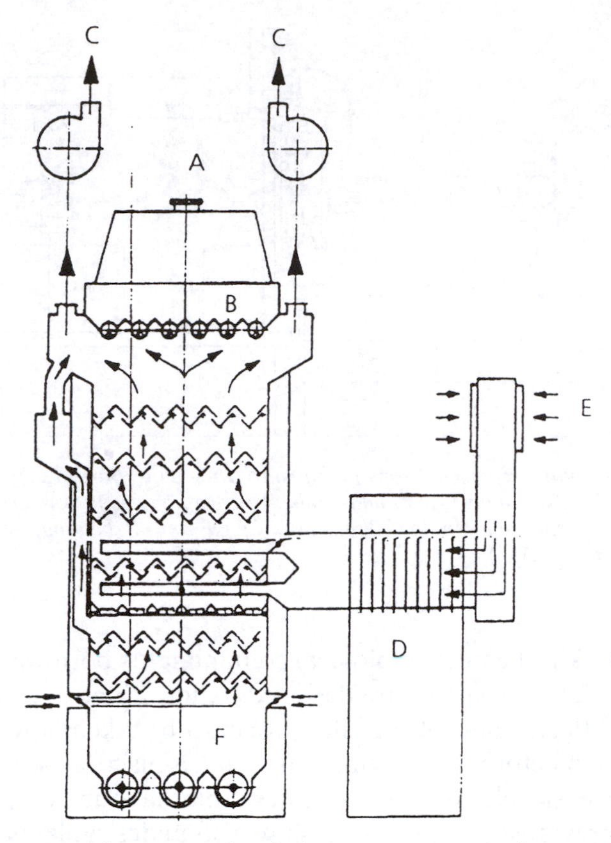

Figure 3.6 *Diagram of continuous bean/nib roasting system (Lehmann Maschinenfabrik GmbH, Germany).* (A) *Product feed;* (B) *feed rollers;* (C) *exhaust air fan;* (D) air heater; (E) *air filter;* (F) *extraction screw* (Heemskerk[1])

cocoa from becoming too hot. This process may take as little as 1 or 2 minutes.

Chemical Changes During Roasting

Unroasted beans usually taste very astringent and bitter. The high temperatures and drying during roasting remove many of the volatile acids, especially ethanoic acid, and makes the nibs, or beans, taste less acidic. The less volatile acids, such as ethanedioic (oxalic) and lactic, remain largely unchanged by the roasting process.

Maillard Reaction

This reaction, also called *non-enzymic browning*, is important to food quality throughout the food industry and gives products their colours and flavours when they are baked, toasted or roasted. It is an extremely complicated reaction involving many low molecular weight components with hundreds of different reactions and intermediate products. These intermediates may or may not have a flavour of their own. Some act as catalysts for other reactions, whereas others may stop a particular series of reactions taking place.

Heat is required for the reaction to take place at a significant rate. Its products can be seen and tasted when food is burned on to a pan, the contents of which have not been stirred well enough. Water must also be present together with a reducing sugar, such as glucose and an amino acid, peptide or protein. In the case of cocoa, about 12–15% of protein is present before fermentation. But the heat and acidic conditions during this treatment break much of this down into amino acid, thereby forming some of the precursors, which are converted by roasting into chocolate flavour.

The principal pathways of the Maillard reaction are shown in Figure 3.7. All the reactions take place at any $pH > 3$, but the actual pH does alter the probability of their taking place. The reactions on the left hand side are largely flavour formation routes. The sugar splits up into smaller carbon chains. (C_1, C_2 *etc.* denote the length of the chains). The key intermediates 1DH, 3DH and 4DH are 1-, 3- and 4-deoxyhexosuloses respectively and are dicarbonyl compounds. The pathways on the right hand side are often involved in colour rather than flavour production.

The Strecker reaction involves the formation of aldehydes, some of which form part of chocolate flavour, from the amino acids, which are

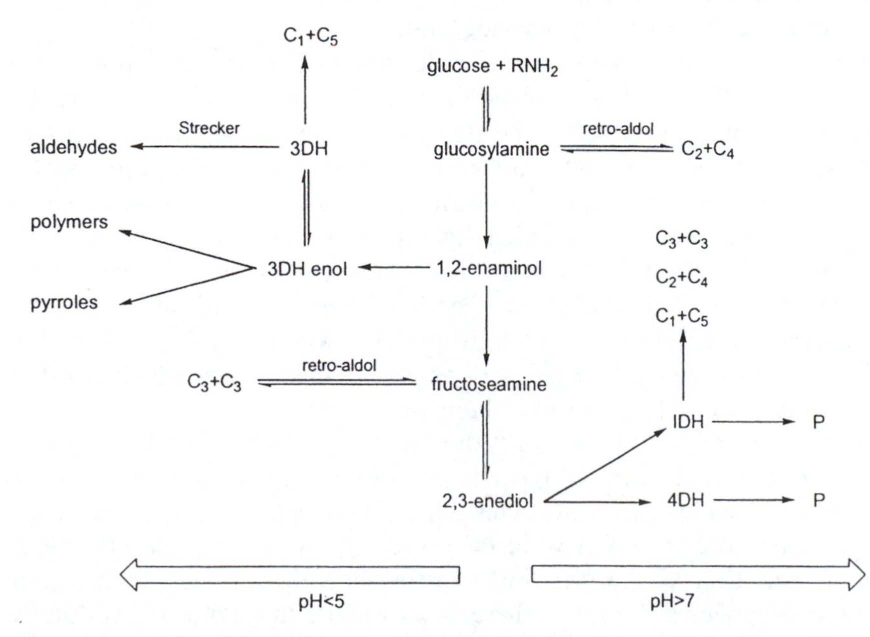

Figure 3.7 *Model of the Maillard reaction*

largely tasteless and odourless. One of these reactions involves the reaction of an amino acid called glycine with glyoxal (a 1,2-dioxo compound). These eventually form pyrazines, which are heterocyclic compounds with two nitrogen atoms in a six-membered ring. The amounts of the different pyrazines formed depends very much upon the temperature and time of the roasting reaction. Their measurement has in fact been used as a method of determining the degree of roast of cocoa liquor (see Chapter 8).

The characteristic smell of chocolate can also be produced by the reaction of amino acids such as leucine, threonine and glutamine with glucose, when heated to about 100 °C. Higher temperatures will produce a much more penetrating/pungent smell.

GRINDING COCOA NIB

There are two aims when grinding the cocoa nib. The first is to make the cocoa particles small enough so that they can be made into chocolate. There is further grinding during the later chocolate making process, so it is not necessary to mill the nib very fine at this stage. The second more important reason is to remove as much fat as possible from the cells within the cotyledons. The fat is needed to help the chocolate flow, both when making the sweets and also when it melts in the mouth. The fat is also the most expensive major ingredient in chocolate, so economically it is necessary to make the most use of all the fat present. The fat is contained in cells which are on average between 20 and 30 microns (μm) long and from 5 to 10 microns wide/high.

Figure 3.8 shows sections through two beans as viewed under a microscope. The fat has been stained so that it appears dark. When moisture is present with fat, the two cannot mix, but can exist in two forms, *i.e.* a water in oil emulsion, with oil/fat surrounding water droplets or an oil in water emulsion in which the oil/fat is in small droplets. Their stability is aided by types of phosphoglycerides, also known as phospholipids, which form a surface between the water and fat. These are called emulsifiers (see also Chapter 5). Within the cocoa nib there is an emulsifier called lecithin. The two types of emulsions can exist within the cells of cocoa, with Figure 3.8a showing a largely oil in water emulsion and Figure 3.8b the reverse phase.

The aim of the milling is to get the fat from within the cells, so that it can coat the solid non-fat particles within the chocolate. Fat can be released by breaking the cells open. There is more fat within the cell than is necessary to coat any new surfaces created by tearing open the cells. This means that as the grinding reduces the cell size the cocoa liquor produced becomes thinner – there is more free fat around. Eventually there is no more fat to be released, and further milling only creates new

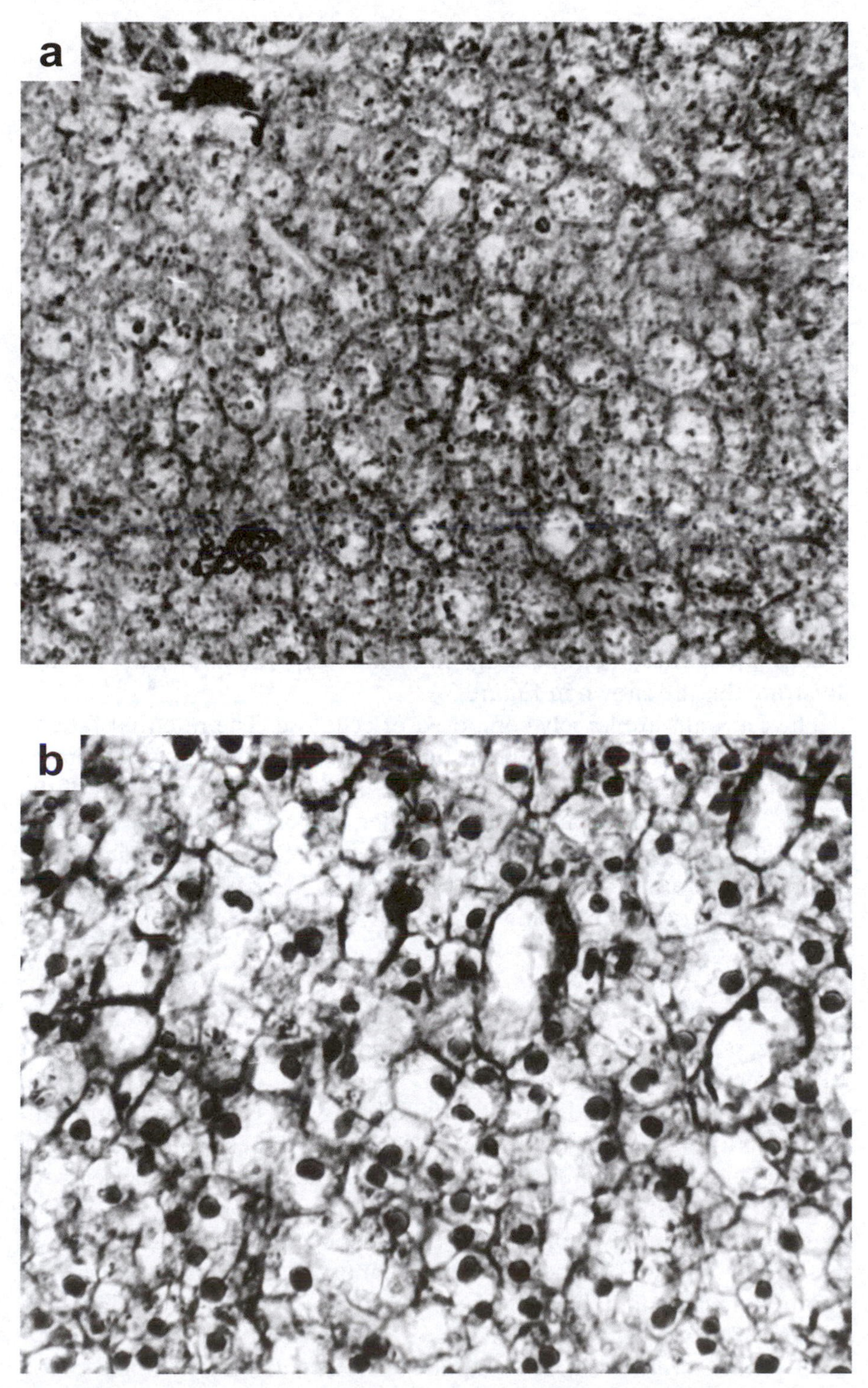

Figure 3.8 *Sections through cocoa beans as seen through a microscope (fat is stained black):* (a) *oil in water emulsion;* (b) *mainly water in oil emulsion*

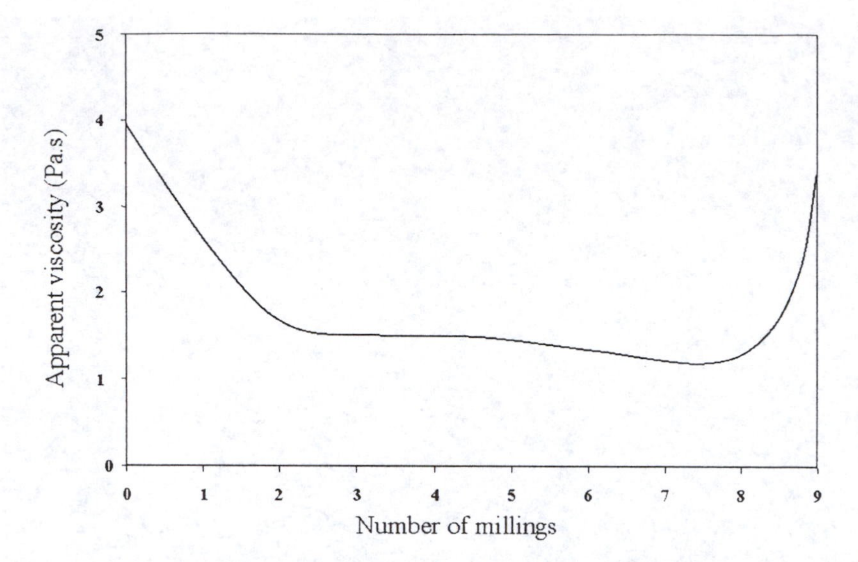

Figure 3.9 *The viscosity of cocoa liquor ground to different finenesses*

surfaces to be coated with fat as it tears apart the cells. This makes the cocoa liquor become thick again. Some experimental figures that illustrate this are shown in Figure 3.9.

The cell walls are largely composed of cellulose. The rate that fat can pass through the cellulose depends upon the amount of moisture that is present (see Project 3 in Chapter 10). When fat is being pressed out of the cocoa, for the production of cocoa powder, this extraction may be aided by the addition of some moisture to the liquor. When grinding, however, as was noted for liquor roasting, it is better to have a lower moisture to enable the cocoa liquor to flow more easily.

Cocoa Mills

It is necessary to grind the nib from a maximum particle size of about half a centimetre down to less than 30 microns. This means that the particles must be ground down by 100 times. Most milling machines can only operate efficiently if the reduction is 10 times, so at least two grinding stages are required. In addition some mills work better with hard material, whereas others will only work with liquids. This means that cocoa is normally ground twice, initially by an impacting mill, which melts the fat and produces a liquid containing big particles several hundred microns in diameter. The second mill is often a ball mill, which will only work with a liquid, or a disc mill, based on the original corn mills, which will work with liquid or solid material. Milled cocoa particles include cocoa starch, which make up about 7% by weight of the cocoa liquor. This has a particle size of between 2 and 12.5 microns and so is not destroyed by the milling process. About 10% of liquor is made up of cellulose and a slightly larger percentage is protein.

If the factory is wanting to press the cocoa liquor to make cocoa powder, the cocoa liquor is not normally ground so finely as when the cocoa liquor is used for chocolate. This is because very finely ground cocoa liquor particles will block up the filters in the cocoa presses and make it more difficult to remove the cocoa butter. In chocolate making, however, it is advantageous to have as much cocoa butter as possible released from the cells.

Impact Mills

Impact mills work by hitting the cocoa nibs with fast moving pins or hammers. Sometimes the particles are hit against sieves or screens. The cocoa butter melts due to the heat from the impact and from the mill itself and any free fat together with the smaller particles pass through the sieve. The larger particles remain on the inside until broken by the next series of pins or hammers.

Disc Mills

Disc mills often consist of three pairs of carborundum discs (see Figure 3.10). The cocoa liquor or nibs are fed into the centre of the top set of discs, where one disc is rotating but the other one is stationary. The discs are pressured together and the cocoa mass is forced through them, by centrifugal force to the outside. The high shear destroys a lot of the particles releasing a lot of the fat. The cocoa liquor then runs down a chute to the centre of the centre set of discs and then finally onto the third set.

Figure 3.10 *Schematic diagram of a triple disc mill* (Lehmann)

Ball Mills

The majority of the world's cocoa is ground using ball mills, which can only grind liquids and so are normally preceded by impact mills. The mills contain large numbers of balls, which are made to bounce against each other, either by tumbling them by rotating the wall of their container or by a centrally rotating shaft with rods placed at intervals at right angles, as is illustrated in Figure 3.11. The balls impact and rotate (Figure 3.12) and any particles caught between then are broken by the crushing or pulled apart due to the shearing of the rotating action. The

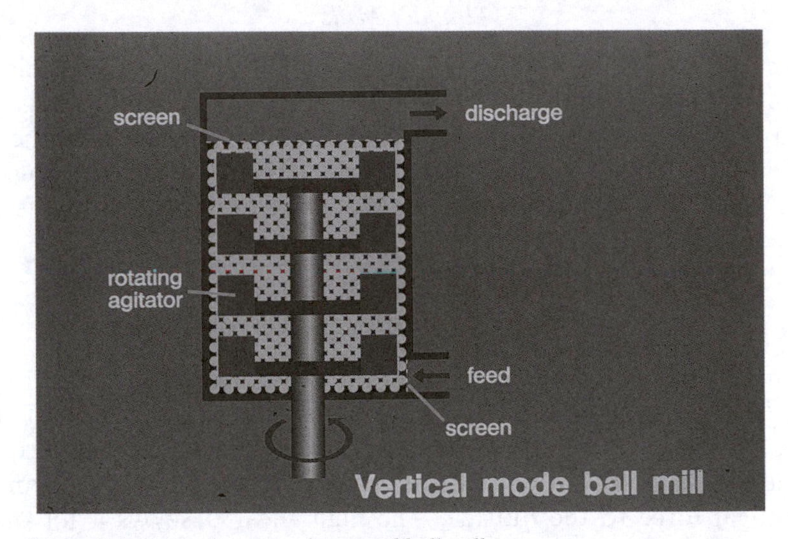

Figure 3.11 *Schematic diagram of a stirred ball mill*

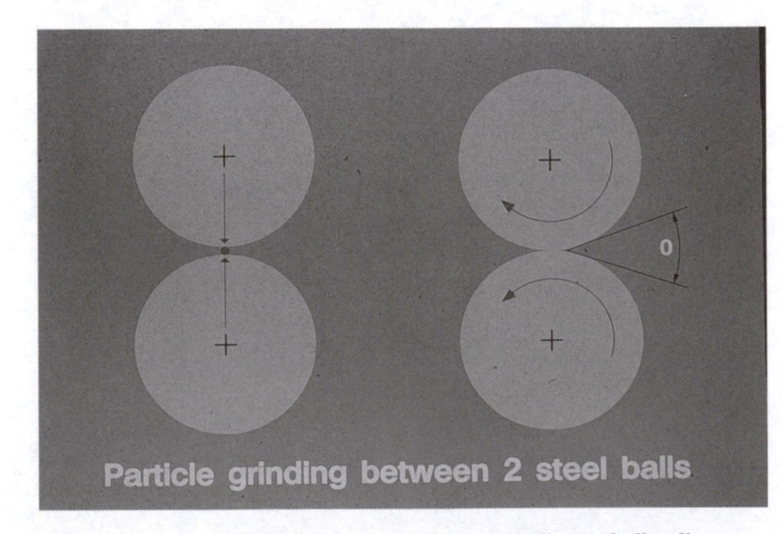

Figure 3.12 *Illustration of the grinding action between balls in a ball mill*

smaller particles move more quickly in the fat as it is pushed away by the moving balls, but the bigger ones are more likely to be milled as they move more slowly. With coarse cocoa liquor containing particles several hundred microns in diameter, the balls may be as large as 15 mm. Where it is desired to have a finer liquor, a series of ball mills may be used, each one containing smaller balls, down to as small as 2 mm. More small balls will fit in the same space, so the chance of a particle being caught between two of them greatly increases. The agitator speed is also increased with the smaller balls.

Sieves are used on the outlet to the mills to stop the balls damaging machinery during the next stages of chocolate making. The balls wear and are replaced at regular intervals. Magnets in the outlet pipes trap any pieces of metal that come out of the mill.

COCOA BUTTER AND COCOA POWDER PRODUCTION

Alkalising (Dutching)

Most of the cocoa liquor that is used to produce cocoa powder is alkalised, whereas very little liquor that is used to make chocolate is treated in this way. The alkalising process was developed in the Netherlands in the 19th century, which is why it is also known as the Dutching process. The reason for doing this was to make the powder less likely to agglomerate or sink to the bottom, when it was added to milk or a water based drink. The actual ability of the alkali to do this is not entirely certain, but it does affect both the colour and the flavour.

A solution of alkali, typically potassium carbonate, is normally added to the cocoa nib before roasting, although the cocoa liquor itself, or even the powder, can also be treated. Care must be taken not to add too much alkali. This is because the cocoa butter molecule is composed of three acids attached to a glycerol backbone (Chapter 6). These acids may react with the alkali to produce soapy flavours. To overcome this, small amounts of ethanoic or tartaric acid may be added after alkalisation, in order to lower the pH. For certain types of beans which taste very acidic, due to the presence of ethanoic and other acids in the cocoa, a mild alkalisation to neutralise these acids can be very beneficial to the eventual cocoa/chocolate flavour.

The reason for the change in colour is due to reactions based on a class of chemicals found in cocoa called tannins (polyhydroxyphenols). These are made up of epicatechin molecules (Figure 3.13), which during the different fermenting, drying and roasting stages may join together, oxidise or react with other chemicals within the cocoa. This increases the number of colour giving molecules and makes the cocoa much

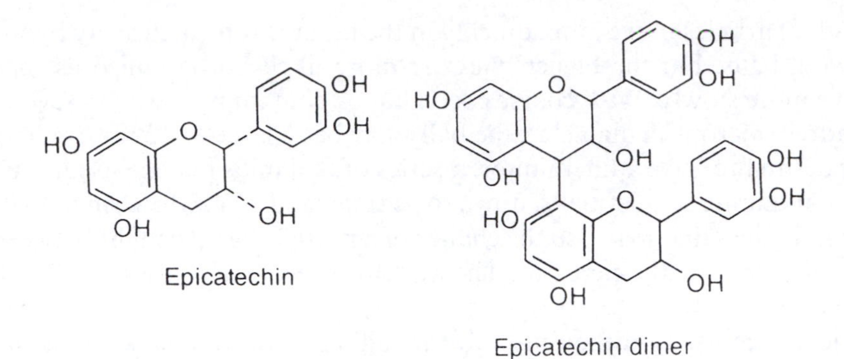

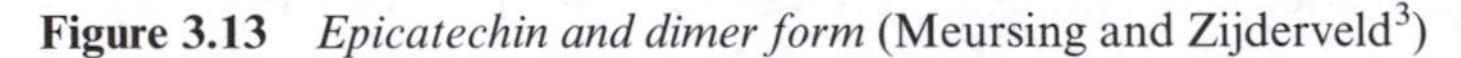
Figure 3.13 *Epicatechin and dimer form* (Meursing and Zijderveld[3])

darker. By carefully adjusting the pH, moisture, roasting temperatures and times, it is possible to produce a wide variety of colours.

Cocoa Butter

The highest quality cocoa butter is obtained by pressing cocoa liquor in a horizontal press of the type shown in Figure 3.14. The top section (4) consists of a series of pots, the base of each being a stainless steel mesh. Hot cocoa liquor is fed into the pot, which is then pressed by a steel ram operating at a pressure of about 40–50 MPa. The nib initially contains about 55% of cocoa butter and this pressure is capable of pushing more than half of it through the sieve, where it flows down a pipe to the weighing point. The pressure is initially applied slowly, to prevent the formation of hard layers that would stop further fat from coming out. It is possible for the operator to increase the pressure until the required amount of cocoa butter has been removed. This leaves a hard material in the pots, which contains between 8% and 24% of fat, depending upon

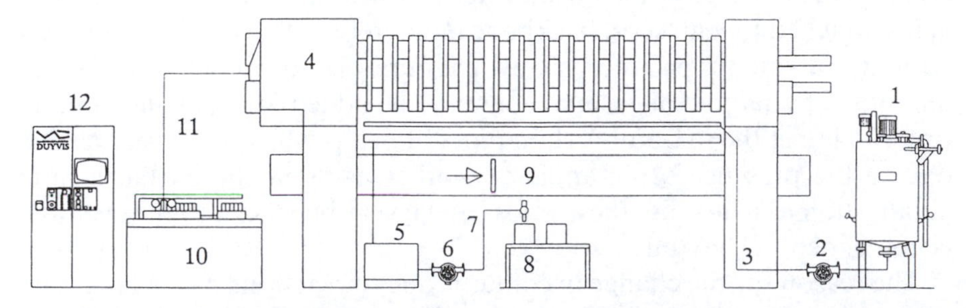

Figure 3.14 *Schematic representation of the operating principles of a horizontal cocoa butter press.* (1) *Cocoa liquor conditioning tank;* (2) *pump;* (3) *pipe for cocoa liquor;* (4) *hydraulic cocoa press;* (5) *cocoa butter scales;* (6) *cocoa butter pump;* (7) *cocoa butter pipe;* (8) *cocoa butter blocking off;* (9) *cocoa cake pushing conveyor;* (10) *hydraulic pumping unit;* (11) *hydraulic pipe;* (12) *control panel* (Meursing and Zijderveld[3])

the type of powder being produced. The hard round discs of material are known as cocoa press cake. When the pots are opened, these are automatically ejected and fall onto a conveyor belt (9).

A lower quality butter is obtained by a continuous expeller process which presses the fat out of whole cocoa beans, which includes the shell. Very often these beans are not of an adequate standard to make chocolate, in that they may not be properly fermented or may have too acidic a flavour. The shell contains some fats that are not cocoa butter. These too are pressed out and get mixed with the cocoa butter. This is detrimental to the hardness and setting properties of the cocoa butter because of its eutectic effect (see Chapter 5). Very often this cocoa butter is cloudy and must be filtered.

Where the press cake is produced from whole or inferior beans, it has little value and is normally used as animal feed. Alternatively the remaining fat can be removed by solvent or supercritical fluid extraction.

When buying cocoa butter, a series of specifications are given to the producer. These include a maximum free fatty acid content (usually 1.75%). This is where acids that have come free from the glycerol 'backbone' of the triglyceride. These will upset the setting properties of the chocolate. Also there is a maximum saponification value (0.5%), to ensure that it does not give a soapy flavour. [The saponification value is the number of mg of potassium hydroxide (KOH) required to react with one gram of the cocoa butter.]

Pure pressed cocoa butter has a flavour, which will become part of the whole chocolate. For some products, in particular white chocolate, this flavour is regarded as unpleasant. In this case de-odorised cocoa butter is used. This is often produced by steam distilling the cocoa butter under vacuum.

Cocoa Powder

Cocoa powder is produced by milling the cocoa press cake. Once this leaves the presses it is broken into pieces less than 3 cm in diameter between two spiked rollers rotating in opposite directions. A cooled pin mill is then used to finely grind the powder. This must then be strongly cooled as it is transported in an air stream through a long pipe to the packing area. Most of the fat is still liquid after the mill, and this must be solidified before packing in order to prevent the powder from sticking together. It is then collected in a cyclone separator with the finer particles being removed by a filter system.

Most powder is produced with a fat content of 20–22%. Lower fat ranges are available, *e.g.* 15–17% or 10–12%. A fat-free powder is produced, but this can not be legally called cocoa powder.

The powder can be mixed with other fats to produce chocolate flavoured (compound) coatings or to produce cake mixes and fillings *etc.* A very large amount is used to make chocolate drinks. These are made from sugar, cocoa powder and lecithin. The lecithin can be added to the broken press cake and then milled with it. This makes sure that the lecithin is closely bound to the cocoa, especially the fat. Lecithin acts as an emulsifier, and forms a boundary between the fat and other cocoa particles and the water when the drink is made. This helps these particles to disperse throughout the water rather than forming lumps.

REFERENCES

1. R.F.M. Heemskerk, 'Cleaning, Roasting and Winnowing', in S.T. Beckett (ed.), 'Industrial Chocolate Manufacture and Use', 3rd Edition, Blackwell, Oxford, UK, 1999.
2. B. Wedzicha, 'Modelling to Improve Browning in Food', 46th Technology Conference, BCCCA, London, 1999.
3. E.H. Meursing and J.A. Zijderveld, 'Cocoa Mass, Cocoa Butter and Cocoa Powder', in S.T. Beckett (ed.), 'Industrial Chocolate Manufacture and Use', 3rd Edition, Blackwell, Oxford, UK, 1999.

Chapter 4

Liquid Chocolate Making

Most people think of chocolate as a solid, because this is how they buy and eat it. To the chocolate maker, however, it is normally a liquid and is only solidified just before it is ready to be packed and sent to the warehouse or shop.

It has already been noted that dark chocolate is made mainly from sugar, cocoa nib and cocoa butter. The basic ingredients for a typical dark chocolate, roughly in the proportions that they are present, are shown in Figure 1.2. Similarly the basic recipe for a milk chocolate made from full cream milk powder is shown in Figure 4.1. As can be seen, these particles are relatively large, some being several millimetres in diameter. As with cocoa liquor, these must be milled so that the largest particles are smaller than 30 microns.

The fine particles must then be coated with fat, so that they can flow past one another when the chocolate melts in the mouth. This process takes place in a machine called a conche (see Chapter 1). This machine is

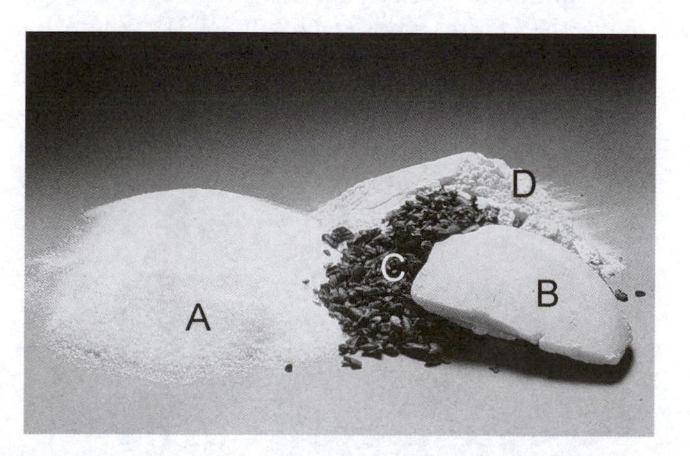

Figure 4.1 *Picture showing the unmilled ingredients used to make milk chocolate:* A, *sugar;* B, *cocoa butter;* C, *cocoa nibs;* D, *full cream milk powder*

Figure 4.2 *Chocolate miss-shape with 'feet'*

the only one specifically designed for the chocolate making process. The mixers are found throughout the food industry, whereas the mills are used to grind many food and non-food products (such as printing inks).

The liquid chocolate is then used to make the final product. This normally takes place by pouring it into a mould or passing it through a curtain of chocolate in a machine known as an enrober (Chapter 7).

How the chocolate flows during these processes is very important in order to produce the correct weight and appearance. The viscosity of the chocolate varies depending upon how vigorously it is stirred or poured and it is what is known as a non-Newtonian liquid (Chapter 5). Its behaviour is often described by two viscosity parameters known as a yield value (related to the energy to start it flowing) and the plastic viscosity (its thickness when it is moving relatively quickly). Both of these parameters must be correct, otherwise waste product may be made. For example 'feet' may form on the base (Figure 4.2) and result in the product being sold at a low price as a mis-shape or there may be holes, which will leave the centre exposed (Figure 4.3). Where the centre is

Figure 4.3 *Chocolate with part of centre uncoated*

moist it will dry out very rapidly, because it is no longer protected by the chocolate and its shelf-life will be very much shorter.

The next chapter looks at the factors that control chocolate flow and which are used by the manufacturer to obtain the correct viscosity for subsequent processing.

CHOCOLATE MILLING

The aim of this process is to make sure that there are no particles in the chocolate that will make it taste gritty, in other words so that there are no particles larger than 30 microns. Another important factor is to ensure that there are not too many very small particles. Unlike cocoa liquor (see Chapter 3, Figure 3.9), which becomes thinner on grinding, chocolate becomes thicker. This is due to the creation of more fine particles, as will be explained in more detail in Chapter 5.

There are two different methods of grinding the chocolate ingredients, namely fine ingredient and combined milling. In the fine ingredient process the solid, non-fat components are milled separately and then added to the cocoa liquor, cocoa butter and other liquid ingredients in the conche. In the combined milling, these ingredients are mixed with the cocoa liquor and some of the other fat before milling takes place. The two processes are likely to give a different flavour, as the sugar will pick up many of the aromas in the mill where it is being ground and in the latter case there is cocoa in close proximity.

Each process has its own advantages and disadvantages apart from the flavour aspects. The fine ingredient process is better able to control the number of fine particles, but the particles are largely fat free at the end of grinding. This means that the fat coating process in the conche takes longer than if many of them were already coated, as is the case for combined milling.

Separate Grinding Mills

The mills are required to reduce sugar and other solid particles by approximately a factor of 100 from millimetres to tens of microns. This is a similar ratio to breaking a brick into pieces the size of sugar grains. When making cocoa liquor it was noted that it was best to reduce particle size in a series of steps, rather than in one operation. This is also true for chocolate ingredients. The hammer/pin mills, of the type used to break the cocoa nib, are very effective in shattering sugar. This is a very brittle material and when hit hard by a fast moving metal hammer or pin will break into a lot of smaller pieces. Milk powder is a lot more elastic and difficult to break and so requires much longer in the mill.

The traditional method of making chocolate was to pass the sugar through such a mill so that the mean particle size was about 100 microns. This was then added to the milk components and cocoa liquor for combined grinding. This is still carried out in some factories, but the process has largely been superseded by a combined grinding process using two sets of roll refiners, as described in the next section.

In order to reduce the particle size still further to the required chocolate particle size, another milling stage is normally required. However, it is possible for the total process to take place in a single machine, called a classifier mill, one type of which is illustrated in Figure 4.4. Within it a series of milling stages can take place before the fine particles leave the mill.

The sugar and dried milk particles are fed into the mill via a chute (1) onto a milling disc (4). This rotates at several thousand times a minute and has metal hammers, wedges or pins at the edge (3), which hit the particles, breaking some and chipping pieces off others. A large volume of air is blown through the mill, entering at (6) and blowing out at (5). This lifts the particles and tries to pull them through the classifier (2). This consists of a rapidly rotating hollow cylinder with slits cut in the side. In fast flowing air, the smaller particles can travel at almost the same speed as the air, but the larger ones are much slower because of their weight and inertia. As the air passes through the slits the smaller ones are able to pass with it and travel out of the mill, where they are collected by cyclones and filter bags. The larger ones, being slower, are hit by the moving bars between the slots and sent back into the breaking

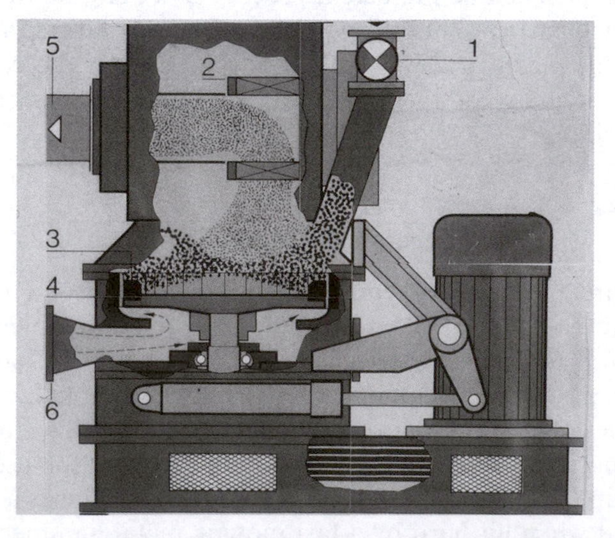

Figure 4.4 *Schematic diagram of a classifier mill as manufactured by Hosakawa Micron.* (1) *Material inlet valve;* (2) *classifier;* (3) *milling hammers;* (4) *milling disc;* (5) *outlet to cyclones and filter bags;* (6) *air inlet* (Ziegler and Hogg[1])

zone again. The larger particle will recirculate for as many times as is required for all of it to be small enough to pass through the classifier. In practice this means that any small sugar particles which are chipped or smashed off the larger crystals pass out of the mill at the first circuit, but some of the sugar and most of the milk will recirculate several times. All the particles within the mill travel at high speed and bump into each other causing a lot of additional breakage.

There are two controls over the particle size that the manufacturer can use to obtain the one needed for the chocolate being made. One is the rate of airflow. If this increases more particles are pulled through the slits and a coarser product is made. If the speed of the classifying cylinder is increased, however, the reverse happens, the bars catch more particles and the product becomes finer. Because the particles pass from the mill as soon as they have been reduced to the required size (unlike combined milling, where all the particles pass through the mill until the largest ones have been destroyed), this type of mill produces relatively fewer finer particles than combined milling.

The mills generate a lot of heat, and this can cause some of the sugar to turn from crystalline into an amorphous form (see Chapter 2). Also any fat present will melt causing the particles to become very sticky and block the pipes. When more than about 12% of fat is present some form of cooling is required. At higher levels still, liquid nitrogen can be fed in with the air and cryogenic grinding carried out. This type of mill is often used to process cocoa powder from press cake, as well as chocolate ingredients.

Combined Milling

In the process used by many modern chocolate factories, the solid particles are milled using roll refiners. The process is shown schematically in Figure 4.5.

Initially the cocoa liquor, granulated sugar and milk components are placed in a heavy duty mixer together with some of the cocoa butter. (For chocolates made with a crumb, most of the solid components are contained within the crumb, which is in the form of a partly milled powder.) It is important to turn this mixture into a uniform paste with the apropriate consistency for the milling to proceed correctly. This paste is then fed into a two roll mill. This consists of two cylinders, placed horizontally side by side, which turn in opposite directions so as to pull the paste into the gap between them. If the paste has the wrong texture it will just form a bridge between the two rolls and the process stops. If it is correct, the pressure and shear in the gap will break some of the particles and also coat some of the newly formed surfaces with fat so that a drier

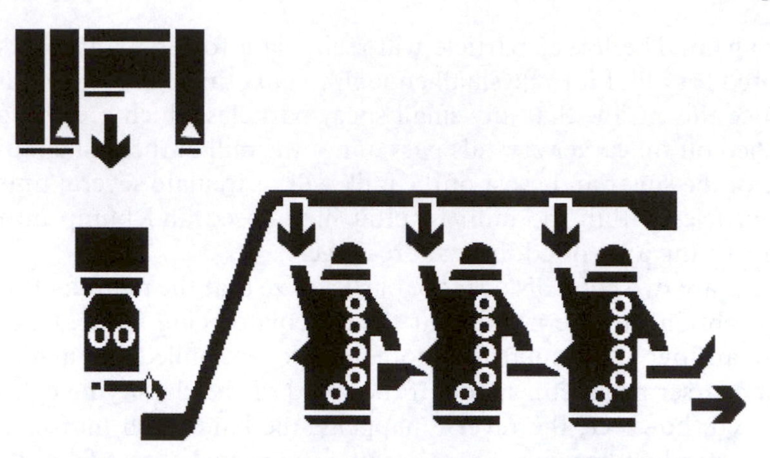

Figure 4.5 *Diagram of two roll and five roll refiner chocolate making plant*

paste is formed, with a maximum particle size of between 100 and 150 microns.

The final grinding takes place on a five roll refiner, similar to the one shown in Figure 4.6, which can be between 75 cm and 2.5 m wide. These take the paste down to a maximum particle size between about 15 and 35 microns. The actual size will depend upon the type of chocolate being made and greatly affects its flow properties as a liquid, as well as the taste and texture in the mouth (see Chapter 5).

The five roll refiner consists of five slightly barrel shaped horizontal cylinders, with four of the cylinders placed one above the other (Figure 4.7). The first, or feed cylinder is placed below the others, but on the side

Figure 4.6 *Picture of five roll refiners*

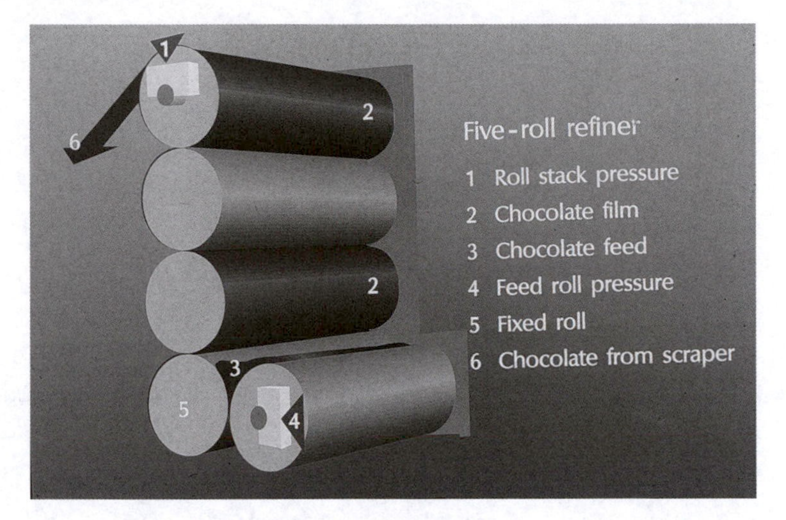

Figure 4.7 *Schematic diagram of five roll refiner*

so that a trough is formed between it and the second cylinder, which will contain the paste from the two roll refiner. Because it has four crushing gaps the five roll machine operates much more slowly than the two roll one, with its single gap. Because of this a two roll machine is normally used in conjunction with several five roll ones.

The cylinders are hollow and can be cooled or heated by water which flows through them. They are also pressed closely together, usually by a hydraulic system. This pressure bends the barrel shape so that it becomes straight and there is a uniform straight gap between the cylinders. A knife blade placed against the back of the fifth cylinder removes the chocolate in the form of flakes or a powder.

The big particles can be broken in many ways. Hitting them hard, as in the hammer mill, can break them into two or more similar sized pieces, or they can chip smaller pieces off the edges. Crushing between two hard surfaces was shown to be one of the ways ball mills operate when milling cocoa liquor (Figure 3.12). Yet another way is to use shear to pull them apart. Shear is related to the difference in speed between two moving surfaces divided by the distance between them. This means that if two surfaces are travelling at very different speeds and are very close together, there is a very high shearing action, which will pull the particles apart. This is in fact the case with a five roll refiner.

Each successive roller is faster than the previous one (Figure 4.7), and because the film of chocolate is attracted to the faster moving surface rather than the slower one, it continues going up the refiner rather than keeping going round and round the bottom one.

This type of machine works by having a continuous film of chocolate from the feed trough (hopper) to the knife. The thickness of the film

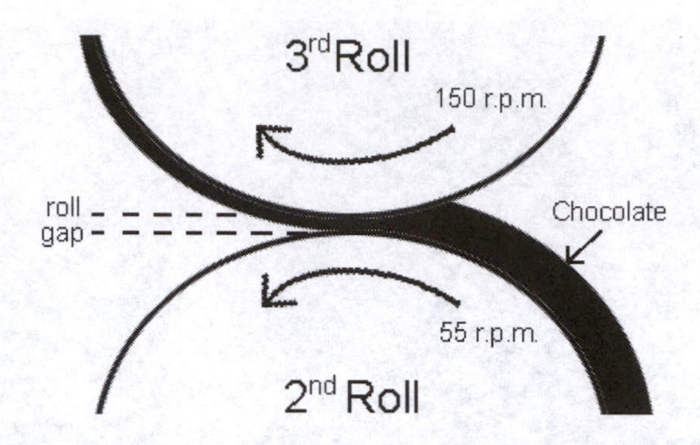

Figure 4.8 *Close-up illustration of the gap between the second and third rolls of a five roll refiner*

depends on the gap between that particular roller and the one below it. What actually takes place is illustrated for the gap between the second and third rollers in Figure 4.8. The lower roller is turning at say 55 rpm and has a film thickness of 100 microns. The second roller is turning at 150 rpm, but the film has to remain continuous. This means that it is stretched out by the higher speed so the thickness is reduced according to the relationship between the speeds of the two rollers, *i.e.* it becomes $100 \times 55/150$ microns thick (= 37 microns). The final fineness therefore depends upon the ratio of the speeds of the different cylinders together with the thickness of the initial film. This latter depends on the outlet width of the slit from the trough, *i.e.* the gap between the first two rolls. Strangely it is affected very little by the pressure, which mainly acts to give a uniform film along the roll.

The temperature also plays an important part in the operation of a roll refiner. This alters the texture/viscosity of the film by changing the flow properties of the fat present. Because the roller surfaces are turning at a relatively high speed there is a centrifugal force on the individual particles trying to throw them away from the machine. The film itself is pulling them on, as long as they remain part of the film. If something goes wrong with the texture of the film, *e.g.* some of the fat sets because it is too cold, the particles become free and the chocolate is thrown from the machine. Temperature control is therefore very important when grinding chocolate.

The shear between the rollers not only breaks the particles, it also coats some of the newly created surfaces with fat. In addition as the breakage occurs, the newly formed surfaces, which are chemically very reactive, are able to pick up the volatile flavour chemicals from the cocoa particles being broken nearby at the same time. This means the chocolate is likely to have a different flavour to one made using the separate grinding process.

CHOCOLATE CONCHING

The chocolate conche was invented by Rudi Lindt in Switzerland in 1878 and was named after the shell, which it resembled in shape. He said that it helped make his chocolate smoother and that it modified the taste. When the conche was invented the ability to mill chocolate was poor, so it is possible that particles were broken in his conche, which made it smoother. Today, however, the grinding systems are very efficient and almost no further breakage takes place, other than if particles are loosely held together as agglomerates. The conche still does change the flavour of the chocolate and also the way it melts in the mouth. Also, of critical importance to the chocolate manufacturer, it determines the final viscosity of the liquid chocolate before it is used to make the final products.

The conching process is in fact two distinct processes which take place within the same machine. The first is flavour development. The fermentation and roasting processes produce the flavour components required to give chocolate its pleasant taste, but they also result in some undesirable astringent/acidic ones that it is necessary to remove. In addition some chocolates need further flavour development, for example for some purposes an enhanced cooked flavour is desirable.

The second is to turn the chocolate from a powder, flaky or thick dry paste into a free flowing liquid that can be used to make the final products. This involves coating the surfaces of the solid particles with fat, so that they can slide past one another.

Chemical Changes

During fermentation ethanoic (acetic) acid and to a lesser extent other short-chain volatile fatty acids such as propanoic and isobutyric (2-methylpropanoic) are formed. However, they have a boiling point above 118 °C, which is considerably higher than the temperature of the chocolate during most conching procedures. Moisture is, however, removed from the chocolate, particularly during the early parts of the process (Figure 4.9), and this may aid the removal of the acids by a type of steam distillation process.

Other workers have noted a large reduction in the amount of phenols during the first few hours of conching. It is not known, however, what if any effect these compounds have upon chocolate flavour. Headspace analysis of the air above a conche has shown that the amount of volatiles decreases by 80% during the first few hours of conching. It is also possible to over-conche. Chocolate that has been processed for too long may have very little flavour at all.

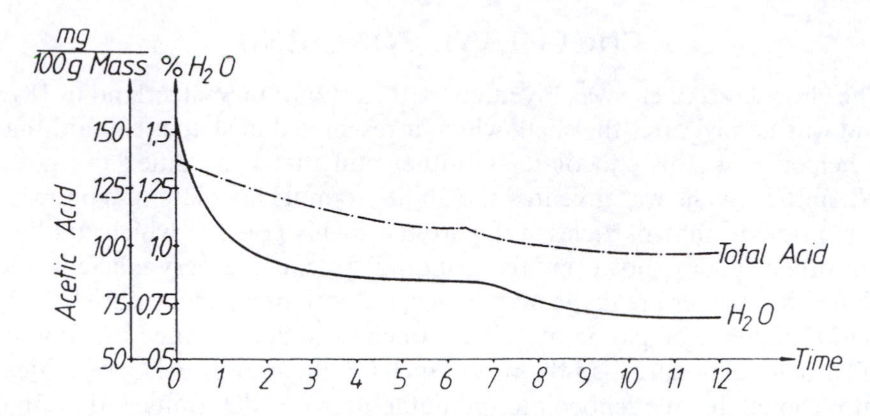

Figure 4.9 *Graph showing the changes in moisture and acidity during a conche cycle (time in h)* (Beckett[2])

The chocolate flavour depends upon the time and temperature used, in general a higher temperature means a shorter processing time. Above about 70 °C in milk chocolate, however, cooked flavour changes start occurring. Some manufacturers use temperatures above 100 °C to try to promote some Maillard type flavours. Because there is very little water present, these flavours are not as strong as those developed during high temperature milk drying or the crumb making process. In some milky flavoured chocolates these Maillard flavours must be avoided so the conching temperatures must be kept below 50 °C. This is also true for some sugar-free chocolates containing sugar alcohols. Here higher temperatures melt the crystals and they later resolidify in gritty agglomerates.

Viscosity Reduction

This process is essentially one of coating the particles with fat. As with grinding, shear is an important factor. Figure 4.10 illustrates some particles at rest and then under shear.

The shear in this case is once again related to the difference in velocity of the surrounding walls divided by the distance between them and we can define a shear rate as:

$$\text{shear rate} = (v_1 + v_2)/h \tag{4.1}$$

Once again high speeds or narrow gaps have a bigger effect upon the particles between them. What we are trying to do is to put fat over the surface, which is a bit like trying to butter bread, where the fat between the knife and the bread is subjected to a high shear and so is forced in a thin layer over the surface.

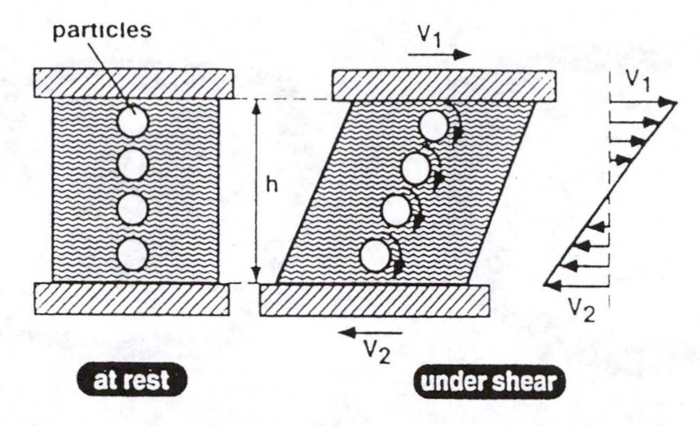

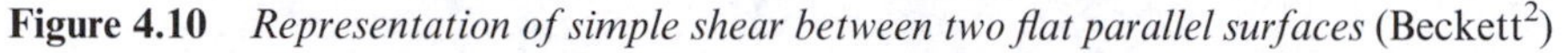
Figure 4.10 *Representation of simple shear between two flat parallel surfaces* (Beckett[2])

In chocolate making this is particularly important because it is desirable to make the chocolate flow as well as possible (have the lowest possible viscosity) at any particular fat content. Figure 4.11 illustrates the same chocolate (identical fat content) processed at different shear rates. As can be seen all reach different equilibrium viscosity readings and they will retain these for however long they are sheared. This means that within the normal conching shear rates used, the higher we can shear a chocolate the thinner it will be. This, however, requires very large motors and a lot of energy and there is a practical limit to the shear that can be developed.

There are two approaches to reducing the chocolate viscosity. One is to have a very large stirred tank, where only a small amount of the chocolate is being sheared at any one moment. Because there is a lot of chocolate in the tank it is possible for the chocolate to be inside the tank for several hours and there still to be a throughput in the region of tonnes per hour. The other is to highly shear a few kilograms at a time in a

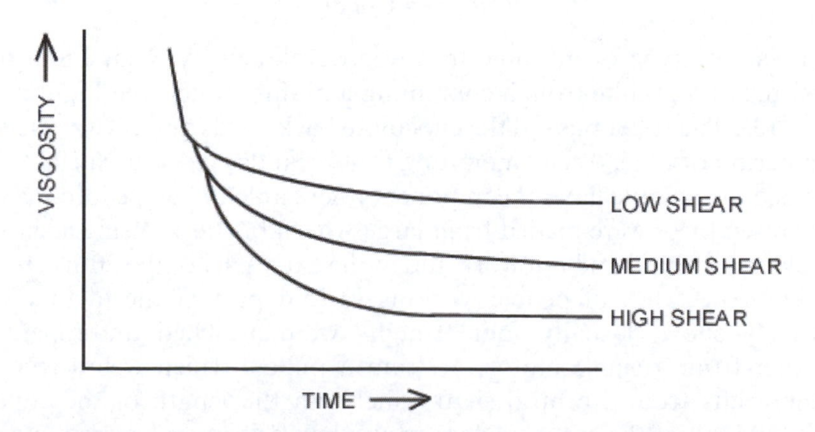

Figure 4.11 *Illustration of the change of viscosity with time for conches with different shearing actions* (Beckett[2])

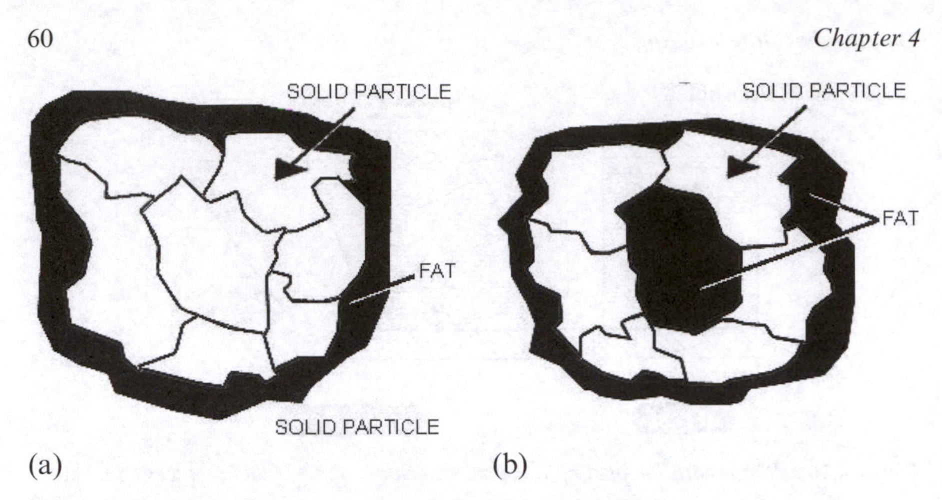

Figure 4.12 (a) *Agglomerate composed of solid particles;* (b) *fat containing agglomerate*

continuous processor. Because of the small amount inside the machine it can only stay there for a fraction of a minute to give the same throughput as a large conche.

Another thing that this process does is to break up groups of particles that are loosely stuck together. These agglomerates can take two forms. In one of them (Figure 4.12a) there is no fat within it, so the breakage just gives new surfaces that have to be coated with fat, so the viscosity is increased. In the other (Figure 4.12b), there is fat in the middle that is surrounded by the solid particles and so does not aid the over flowability of the chocolate. Once this is broken, however, more fat is released than is necessary to coat the new surfaces, so the overall viscosity is reduced.

Conching Machines

The Long Conche

This was the type of machine that was developed by Rudi Lindt and consisted of a granite trough containing a granite roller (see Figures 1.3 and 4.13). The roller pushed the chocolate backwards and forwards for a long period of time, often amounting to several days. As the surface was changed the volatile flavour components were able to escape into the air. Very often these were loaded from large wheelable tanks using a shovel. This was very hard and hot work and in the early part of the 20th century the life expectancy of people working in that part of the factory was relatively short. Usually four troughs were attached and operated together from a single motor, or from a pulley, which was driven by leather belts from a central shaft which ran the length of the conche rooms. Much of the early machinery was belt driven and some factories had saddler's shops to maintain the leather.

Figure 4.13 *Picture of a long conche being filled*

The density of the powder that is fed into the conche is less than half that of the finished chocolate, so the troughs appear relatively empty at the end of the process, even though they may have been nearly overflowing at the beginning.

These conches have poor temperature control, a high energy consumption together with a relatively small capacity and so have almost entirely been replaced by more modern designs.

Rotary Conches

This type of conche is so named because the mixing elements rotate within the tank body, which forms the outside of the conche. Many of the earlier designs (Figure 4.14) were round with a vertical central shaft which drove the mixing or scraping arms. Others had additional mixing elements which performed a planetary motion around the centrally driven arms. Most were open topped to enable the moisture and flavour volatiles to escape. As with the long conche they tended to have poor temperature control and tended to be limited in size to about 1 tonne capacity, although 3 and 5 tonne machines were made. They have therefore been largely superseded by conches with horizontal stirring elements.

A typical modern conche is illustrated in Figure 4.15. Here the tank has three connecting troughs and three stirring arms. As they rotate they smear the chocolate against the side of a temperature controlled wall – the 'buttering bread' action. It then throws it in the air, which enables the volatiles and moisture to escape more easily. The previous forms of conches normally couldn't do this as they tended to compact the chocolate into the bottom of the conche.

Figure 4.14 *Picture of rotary conches*

The ends of the conche arms are wedge shaped. This means that when the chocolate is a thick paste it can operate with the point first. This will be able to cut into the paste and the side of the wedge can smear it against the wall. When the chocolate becomes thin, on the other hand, it will just flow around the wedge and not much mixing or coating of the particles takes place. By reversing the direction of the arms at this stage however, the flat end of the wedge causes much more movement and mixing to take place.

Many of these conches have louvres on top (Figure 4.16), which can be fitted with fans to aid volatile removal, if required. This makes them safer and more hygienic. Conches often process between 5 and 10 tonnes of chocolate in a period of less than 12 hours. The conches are filled automatically from conveyor belts feeding from the milling stage. They also empty automatically through pipes in the base. This is just one more illustration of how in about twenty years the chocolate industry has changed from being a highly labour intensive almost craft industry to a modern, high throughput, large machinery one.

Continuous Low Volume Machines

This type of machine tends to be used to produce a lower viscosity chocolate once the flavour changes have taken place (although twin screw extruders have been used to carry out both processes). A typical liquefying machine is illustrated in Figure 4.17. It contains a central shaft

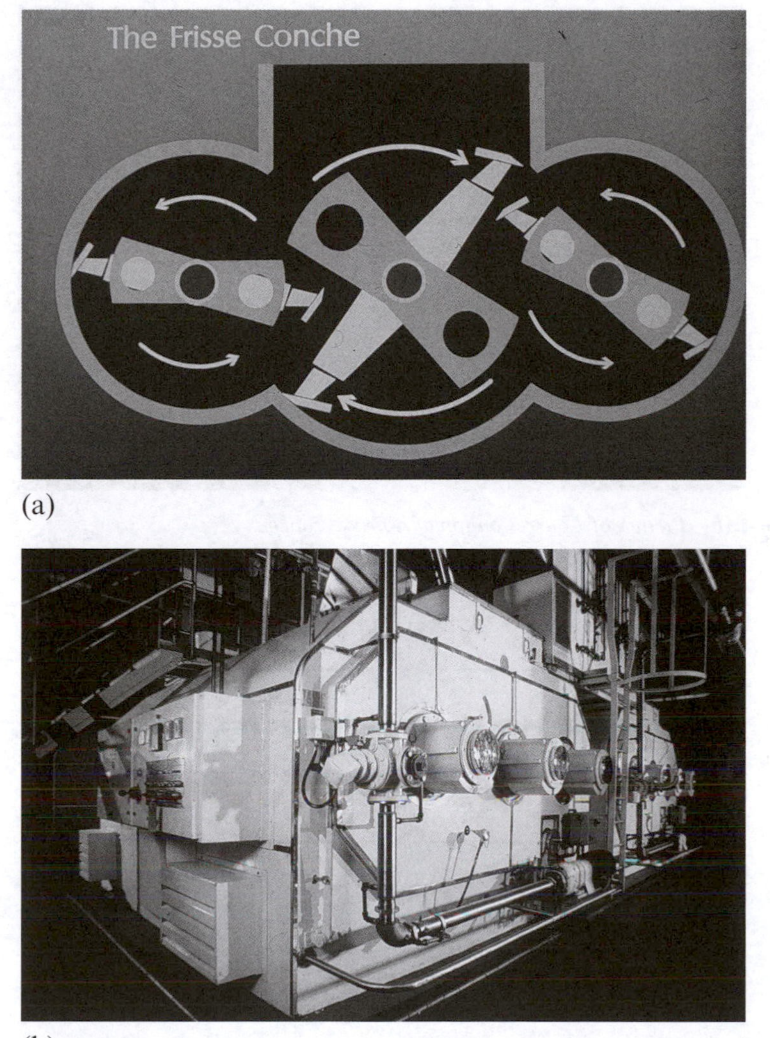

(a)

(b)

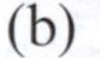

Figure 4.15 *Schematic diagram and picture of a conche manufactured by Frisse, Germany*

with pins attached to it at intervals and which rotates at a high speed. The walls have pins attached which do not move (stators). The gaps between the pins are small and the velocity across this gap is very high, *i.e.* a very high shear rate. As the chocolate is pumped along the pipe it is subjected to this violent action, which causes the fat to coat much of the solid surfaces. The energy input is very high, however, which makes the chocolate become hot. This would alter the flavour of the chocolate were it not for the pipe walls, which are cooled by cold water.

This type of machine can be used after conching or in parallel with it, *i.e.* by having some of the chocolate pumped out, liquefied and then

Figure 4.16 *Picture of Louvres on top of a Frisse conche*

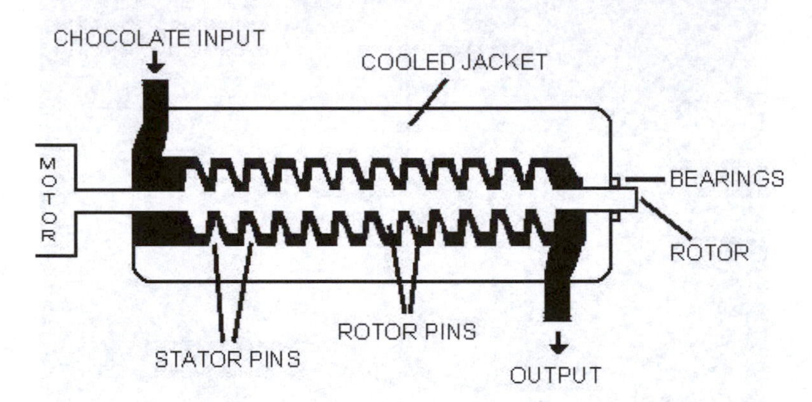

Figure 4.17 *Diagram of an in-line chocolate liquefier*

returned to the conche. In both cases the aim is to produce a thinner chocolate in a shorter time.

The Three Stages of Conching

To get a well processed chocolate it is normally desirable to let it pass through three stages:

(1) dry conching
(2) pasty phase
(3) liquid conching.

In the dry phase the chocolate is still powdery and, for milk chocolate in particular, contains an excess of moisture. This is detrimental to the chocolate's flow properties (see Chapter 5). In addition, when it is removed it is able to take with it some of the undesirable acidic flavours (Figure 4.9). When a lot of the surfaces are still uncoated with fat, it is a lot easier for the moisture to escape. This means that if the chocolate is heated and mixed in the powdery state the whole process can take place more quickly and a lower moisture, thinner chocolate is produced.

As the temperature rises more of the cocoa butter melts and the particles begin to stick together. Sometimes they form into balls several centimetres in diameter, which run around the conche before joining together to form a thick paste. Within the paste there are still a lot of milk and/or sugar particles which are still not coated with fat. When the paste is thick there is a relatively high probability of the shear/smearing action coating them with any fat that is nearby. Once it becomes thin, however, these uncoated particles will just flow out of the way. In order to make a chocolate which flows well in the mouth (when the fat melts due to the body heat) it is necessary to coat as many of these surfaces as possible. This means that the paste should be kept as thick as it is possible for the conche motor to mix, for as long as possible.

The final function of the conche is to ensure that the chocolate has the correct flow properties for the next processing stages. This in turn will depend on the type of coating or moulding machinery being used. The final stage is therefore one in which the final additions of fat and emulsifier (see Chapter 5) are added to the chocolate. This makes it very thin and little further mixing takes place.

The chocolate can then be pumped into storage tanks ready for use. Sometimes it is transported as a liquid in a road tanker to another factory. Alternatively the chocolate can be solidified and stored or transported as blocks or small chips.

REFERENCES

1. G. Ziegler and R. Hogg, 'Particle Size Reduction', in S.T. Beckett (ed.), 'Industrial Chocolate Manufacture and Use', 3rd Edition, Blackwell, Oxford, UK, 1999.
2. S.T. Beckett, 'Conching', in S.T. Beckett (ed.), 'Industrial Chocolate Manufacture and Use', 3rd Edition, Blackwell, Oxford, UK, 1999.

Chapter 5

Controlling the Flow Properties of Liquid Chocolate

The flow properties of liquid chocolate are important to the consumer and the confectionery manufacturer.

Although there are many very sophisticated instruments for measuring viscosity and texture, the human mouth is really far more sensitive than most of these. When someone eats chocolate, the teeth bite through the solid chocolate. This means that the hardness of the solid chocolate is very important. The temperature of the mouth, at about 37 °C, is above the melting point of the fat within it, so the chocolate rapidly melts especially as it is subjected to the intense mixing and shearing of the teeth and tongue. Once it has melted there are two important factors. One is the maximum particle size. As was noted earlier, if there is a significant number of particles larger than 30 microns (0.03 mm) the chocolate will feel gritty on the tongue. In addition a difference of 2–3 microns in maximum particle size, for sizes below 30 microns, can be detected as different levels of smoothness. Chocolates with a maximum particle size of about 20 microns have been sold as having a silky texture. The second factor is the viscosity. This not only affects the way the chocolate runs around the mouth, *i.e.* the texture, but it also changes the taste. This is because the mouth contains three different flavour receptors in different places (see Figure 5.1). The times the solid particles in the chocolate take to reach the receptors depend upon the viscosity. This means that two chocolates made from identical ingredients, but processed to give different viscosities, will taste very different (see Project 15, Chapter 10). Particle size affects viscosity as well as texture, and a milk chocolate which has been milled to say a maximum particle size of 20 microns will have a creamier taste and texture than a 30 micron one.

For the manufacture, the weight control of his products is very important. Chocolate is a relatively expensive food in terms of both ingredients and processing. It is therefore economically important not to

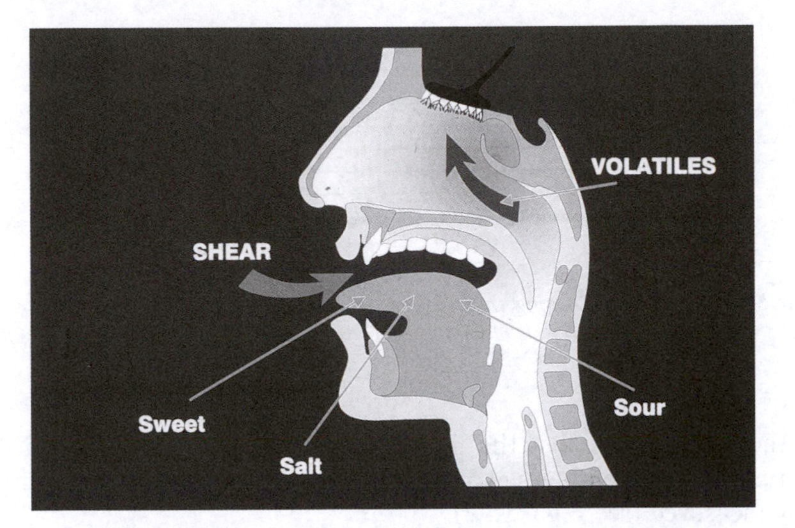

Figure 5.1 *Diagram showing the location of flavour receptors in the human head*

put too much chocolate on a confectionery centre. As will be shown in Chapter 6, the coating methods rely on the liquid chocolate having the correct viscosity. In addition, if the coating is not put on correctly, mis-shapes will be formed (see Figures 4.2 and 4.3), or the centre will be exposed reducing the shelf-life of the product.

VISCOSITY

Viscosity has been described as 'the resistance to motion when stirred or poured'. This is, however, not necessarily a single number. Non-drip paint or tomato ketchup are everyday examples of materials with complex flow properties. The paint tin or the ketchup bottle can be carefully turned upside down with their lid removed, and the contents will initially remain in place. If on the other had they are vigorously stirred or shaken just before being turned, they will pour out as quite thin liquids. So how is viscosity defined?

We can consider viscosity or consistency as an internal friction to movement, when movement is easy (*i.e.* runny materials) there is little friction, but for thick materials friction is high. For this it is useful to reconsider the idea of shear (see Figure 5.2). If the liquid has two flat surfaces each of area A and a distance h apart moving at velocities v_1 and v_2, then from the previous definition we have a shear rate (D) of $(v_1 + v_2)/h$ (equation 4.1). Because velocity has the units of length divided by time and we are dividing this by a length, h, the shear rate must have the units of 1/time and is normally measured in s^{-1} (reciprocal seconds).

The force required to move the top plane relative to the bottom one is called the shear stress (τ). If we plot the shear rate against the shear stress,

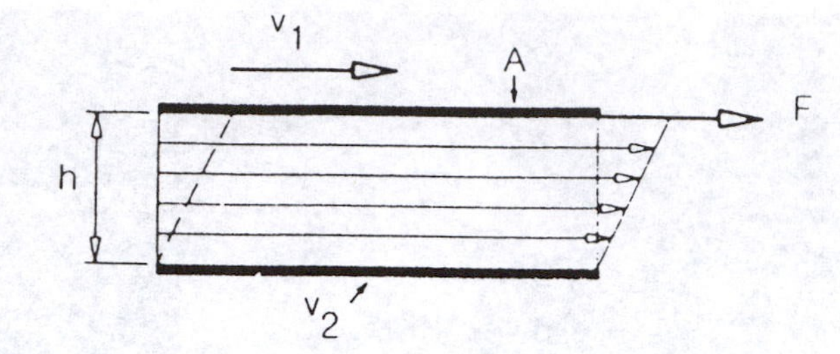

Figure 5.2 *Diagram illustrating shear.* v_1 = *velocity* (cm/s); v_2 = *velocity in lower plane; A = surface area of plane* (cm^2); *F = force; h = distance between the planes* (cm) (Nelson *et al.*[1])

in other words how fast the liquid moves as different forces push it, different curves are obtained, depending upon which type of material is being measured (see Figure 5.3).

Viscosity (η) is defined as the ratio of shear stress to shear rate, *i.e.*

$$\eta = \tau/D \tag{5.1}$$

The units of viscosity are Pascal seconds (Pa s), although the older units of Poise (0.1 Pa s) are also still used. This is illustrated in Figure 5.3 where the viscosity is the gradient of the line, which for liquid number 1 is the same for all shear rates. So if we double the force on this liquid it will move twice as fast. This is known as a Newtonian liquid, as it was first described mathematically by Sir Isaac Newton.

Substances such as golden syrup are Newtonian, but the vast majority of foods are non-Newtonian and follow a wide variety of other curves. Curve number 2 shows a substance which does not move when smaller

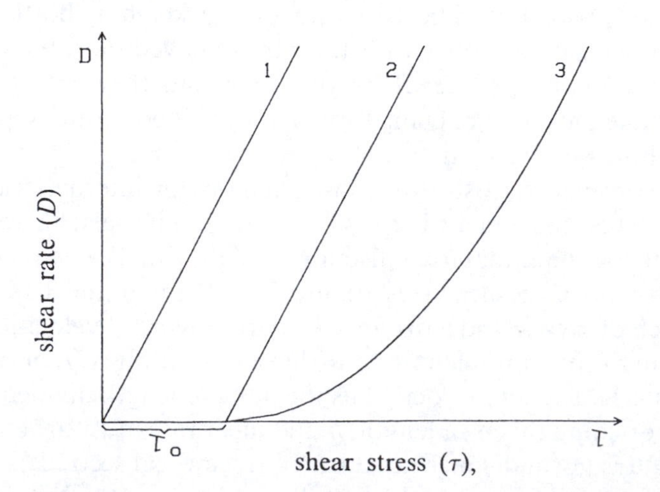

Figure 5.3 *Different types of flow curve:* (1) *Newtonian;* (2) *Bingham;* (3) *pseudoplastic (e.g. chocolate)* (Chevalley[2])

forces are applied, just as was the case when the paint or ketchup was inverted. Once it starts to move, however, it behaves just like a Newtonian liquid. These substances are known as Bingham fluids.

Chocolate has a more complex flow behaviour, however. As with the Bingham liquid it takes a measurable force to start it flowing, but once it does so, the higher the force the thinner it becomes. This is a major problem for the chocolate manufacturer, because the viscosity is not a single number, but has a value which depends on how fast it is flowing. It can in fact be represented best by a flow curve.

If you are in a factory it is not possible to use a curve as a specification for a chocolate viscosity, so this data has to be simplified. The way that this is normally done is to describe the curve by a mathematical equation. Many exist, but the most widely used is the Casson equation, which was originally developed to describe the flow of printing ink. The viscometer (see Chapter 8) measures a few points on the curve and then the equation is used to give two flow parameters: the yield value and the plastic viscosity.

The yield value is related to the energy required to start the chocolate moving. If it is high the chocolate will tend to stand up, which may be required for putting markings on sweets or when producing chocolate morsels for baked cookies. A low yield value is needed to give a thin coating of chocolate over a biscuit.

The plastic viscosity relates to the energy required to keep the chocolate moving once it has started to flow. This is also important in determining coating thickness of chocolate on a sweet and also in determining the size of pumps needed to pump the liquid chocolate.

The remainder of the chapter describes how the chocolate processing and ingredients can be used to adjust the viscosity of the chocolate.

PARTICLE SIZE

Particle Size Distribution Data

In the first part of this chapter it was shown that the viscosity of chocolate could not be described by a single number, but that a minimum of two numbers were needed and to be really accurate, a full curve was required. Particle size is the same. Although, so far, the maximum particle size has been referred to as determining the texture of a chocolate, very few particles are as large as this. Once again the real situation is a curve, known as a particle size distribution and the chocolate manufacturer has to take a summary of this information to aid with processing and quality control.

The particle size distribution can take several forms; two of the most common are illustrated in Figure 5.4. This in fact shows the size

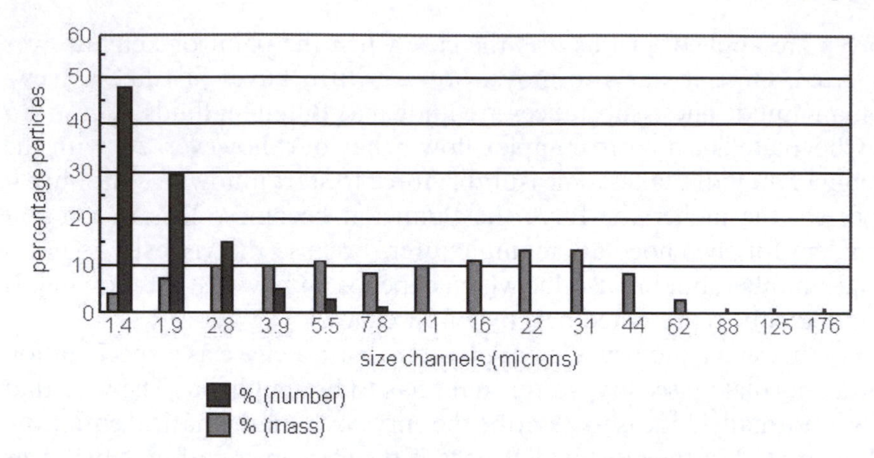

Figure 5.4 *Comparison of mass and number particle size distributions of a chocolate*

distribution of the same chocolate but displayed in two ways. If the solid particles in the chocolate were viewed down a microscope and the diameter of each determined, it would then be possible to determine the proportion of particles within a given size range. This is shown in the number *vs.* size distribution. If on the other hand the individual particles were weighed and the proportion of the total weight that was due to a certain size range was measured, then a distribution like the other one in Figure 5.4 would be obtained. This is a mass *vs.* size distribution. Normally these are measured by determining an estimate of the volume of the solid particles and then assuming a constant density. (See Chapter 8 for details of the type of the instrumentation capable of making these measurements.)

As can be seen the two curves look very different. The mean particle size on the number distribution is about 2 microns, whereas it is almost 10 microns for the mass distribution. The maximum size on the number distribution appears to be about 9 microns, although it is known from the mass distribution that particles as large as 70 microns are present. Why is this so?

The difference is due to the fact that the number distribution is proportional to the particle's diameter or radius (r), whereas the mass relates to its volume, that is the cube of the radius (volume $= 4/3\pi r^3$). A ten micron particle is therefore equivalent in mass to $10 \times 10 \times 10 = 1000$ one micron ones on the mass scale, but the two are equal on the number scale. Although the large ones are present in the sample, on the number graph their proportion is too low to be significant and to be registered.

In chocolate a small number of larger particles will give a gritty texture, so the number distribution will not give the information that is needed. The mass distribution is better, but it is still not obvious what measurement to take to be representative of the large particles, *i.e.* the grittiness of the chocolate. Once again it is a distribution. Particles much larger than 50 microns may be present in the occasional sample, but it would be impractical to sample a lot of the chocolate. In fact the maximum particle size is going to vary very much from sample to sample within a bar of chocolate. The solution that is often used is to record the 90th percentile. That is, the size at which 90% of the particle mass is due to particles with a diameter less than this size. This appears to correlate fairly well with what people actually taste and with measurements made of the biggest particles using a micrometer (Project 6 in Chapter 10).

The Effect of Particle Size upon Viscosity

The largest particles are important for mouth-feel with respect to grittiness, but the smaller ones are more important with respect to chocolate flow properties, in particular the yield value.

The reason for this is because a large amount of fat is required to coat them so that they can move past one another in the liquid chocolate. Figure 5.5 represents a cube of sugar. This has six faces (four sides, the top and bottom), which need coating with fat. If this is broken in half, then there are eight sides to be coated (the two new surfaces are the same size as the old ones). This means that there is a need for 30% more fat, although there is only the same amount of sugar present. As was noted in the previous chapter, it is necessary to reduce the particle diameter by

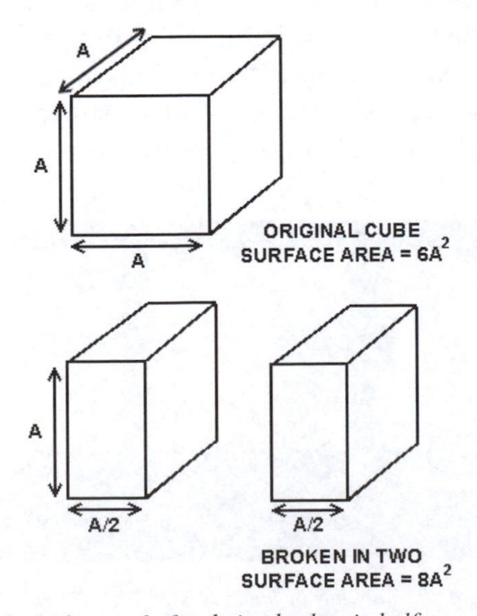

Figure 5.5 *Sugar cube before and after being broken in half*

about 100 times, so a vast new surface is created, which will use up the fat which helps the chocolate to flow.

It is known that we need to mill the sugar below 30 microns, otherwise it would be rough. What would happen if it were possible to create spherical sugar particles all of which were of diameter 29 microns? This would have a minimum surface area to be coated with fat, but would have all the particles small enough to make chocolate. This, however, would give a very thick chocolate. The reason is that the particles would not pack together. As is shown in Figure 5.6, all spherical particles can only fill about 66% of the volume. If we take another size of specially chosen smaller ones to fill in the holes this increases to 86% and a third size will take this to 95%. So the chocolate manufacturer ideally would like to grind his particles so that they pack well together but have a minimum area to coat. At the moment it is only possible to make minor changes to the particle size by adjusting the operating speeds and conditions of the mills being used.

So what parameters have to be measured to determine whether the particle size distribution is a good one for chocolate making? The shape of the particle size distribution curves gives some indication but, as with viscosity measurements, curves are difficult to use to control a manufacturing process. There is, however, one parameter which gives a useful guide as to how the particle size distribution will affect the flow.

Figure 5.7a represents a lump of sugar wrapped up as a present. If it is unwrapped and the minimum amount of paper has been used, we have a measurement of the surface area of the sugar (see Figure 5.7b). If all the 'pieces of paper' from the particles within a defined volume are fitted together, it produces an area per unit volume, also known as a specific surface area. Machines are available to produce this measurement, which is normally expressed in m^2/cm^3. This measurement relates to the radius squared (the surface area of a sphere is $4\pi r^2$). So this measurement in fact is not obtained directly from either the number or

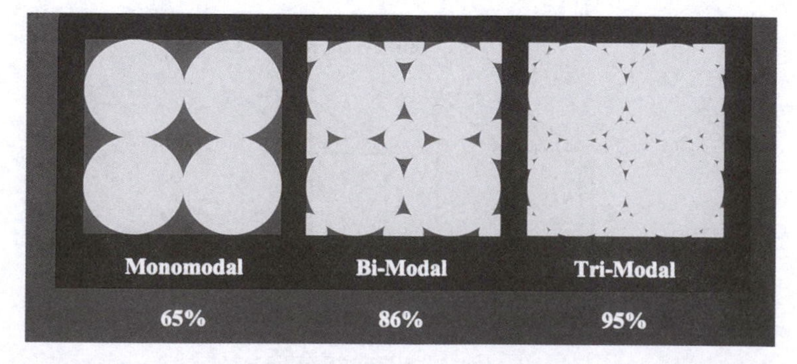

Figure 5.6 *Illustration of the packing of spheres that are:* (1) *all the same size;* (2) *two different sizes;* (3) *three different sizes*

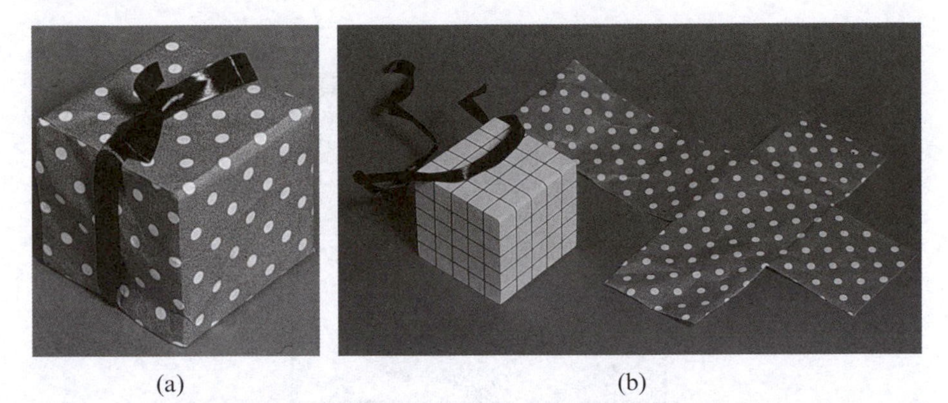
(a) (b)

Figure 5.7 (a) *Wrapped up cube;* (b) *unwrapped cube, showing the minimum area of paper needed to cover it*

the mass distribution. A surface area distribution would lie between the two.

It has been shown that the fine particles make the chocolate thicker by reducing the amount of fat present in the chocolate that can enable the particles to pass by one another. They don't, however, affect both of the flow parameters in the same way. Figure 5.8 shows graphs of the yield value and of the plastic viscosity of the same chocolate that has been milled to different particle sizes. The yield value increases dramatically as the chocolate becomes finer, but the plastic viscosity stays almost unchanged, in fact decreasing slightly at one stage. This is because over half the volume of the chocolate is taken up by solid particles, as is illustrated in Figure 5.9.

When the particles are large there are a limited number of points of

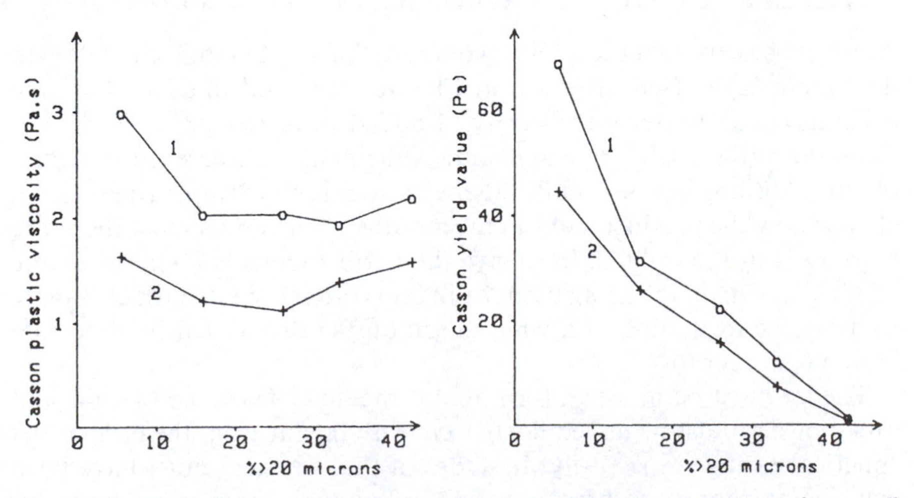

Figure 5.8 *The influence of particle fineness on the viscosity parameters of two milk chocolates:* (1) *30% fat;* (2) *32% fat* (Chevalley[2])

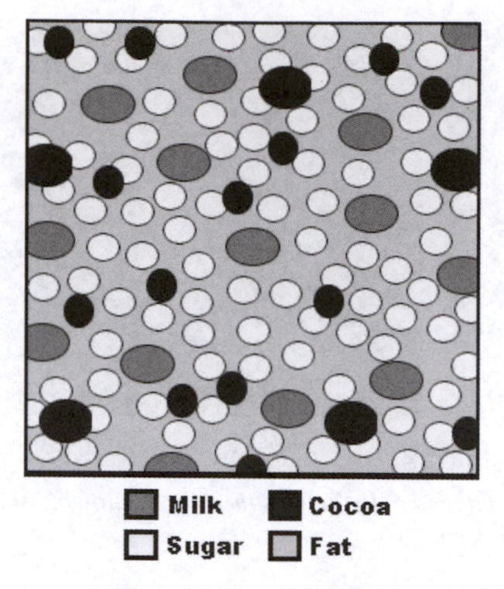

Figure 5.9 *Schematic representation of the solid particles and fat within a milk chocolate*

contact. As the number of particles increases, there are many points where the particles are near one another, so a loose structure is built up. This structure has to be broken before the chocolate will flow, *i.e.* the structure increases the yield value. Once the chocolate begins to flow the structure is broken and the small particles can move along together. There is therefore very little difference between that and big particles moving, so the plastic viscosity remains almost unchanged. The slight decrease is probably due to more fat being released into the system by further grinding of the cocoa liquor or milk powder. As was shown in Chapter 3, cocoa liquor becomes thinner as it becomes finer, in contrast to chocolate.

THE EFFECT OF FAT ADDITIONS UPON VISCOSITY

As would be expected the addition of more liquid fat helps a chocolate to flow more easily (see Project 5 in Chapter 10). Milk fat has the same effect as cocoa butter on viscosity, if added to chocolate at 40 °C, but slows down the setting rate and softens the final chocolate (see Chapter 6). In addition, because milk fat melts at a lower temperature it will change how the product melts in the mouth. The two fats must therefore be present in the right ratio to give the correct chocolate texture in the product in which it is being used. Milk fat is often used in dark chocolate to delay the formation of a white sheen on the surface called chocolate bloom (Chapter 6).

The fat must be in a free form to aid the flow. The cocoa liquor and milk powder must be milled finely to release the fat from the cells or the spherical casing respectively. In addition the conching must have been sufficiently vigorous to break any fat-containing agglomerates.

Most chocolates contain between 25% and 35% of fat, although ice-cream coatings are much higher and some special products like cooking chocolate and vermicelli pieces are lower in fat. The actual level present will depend on the process being used – a certain amount is needed so that the chocolate film remains on the roll refiner and so that the conche motor is not overloaded. In addition it will affect the texture of the finished chocolate, so a high quality tablet of chocolate is likely to have a higher fat content and a lower particle size than a chocolate that is used to coat a biscuit.

The effect of an extra 1% of fat upon the viscosity depends upon the amount that is already there and also which of the viscosity parameters we are considering (see Figure 5.10). Above a fat content of 32% there is very little change in viscosity with any further additions. A 1% increase to a 28% fat content has a really dramatic effect especially on the plastic viscosity, which is almost halved. The change becomes more dramatic at even lower fat contents as chocolates below 23% fat are normally a paste rather than a liquid, but 25% fat chocolates are available on the market.

The effect of fat is proportionately much higher for the plastic viscosity than the yield value (more than 12 × compared with less than 3 × for the samples and ranges illustrated). This is not too surprising as the extra fat will add to the free moving fat that aids particles when they flow past each other. The majority of the fat is 'wetting' fat, which is

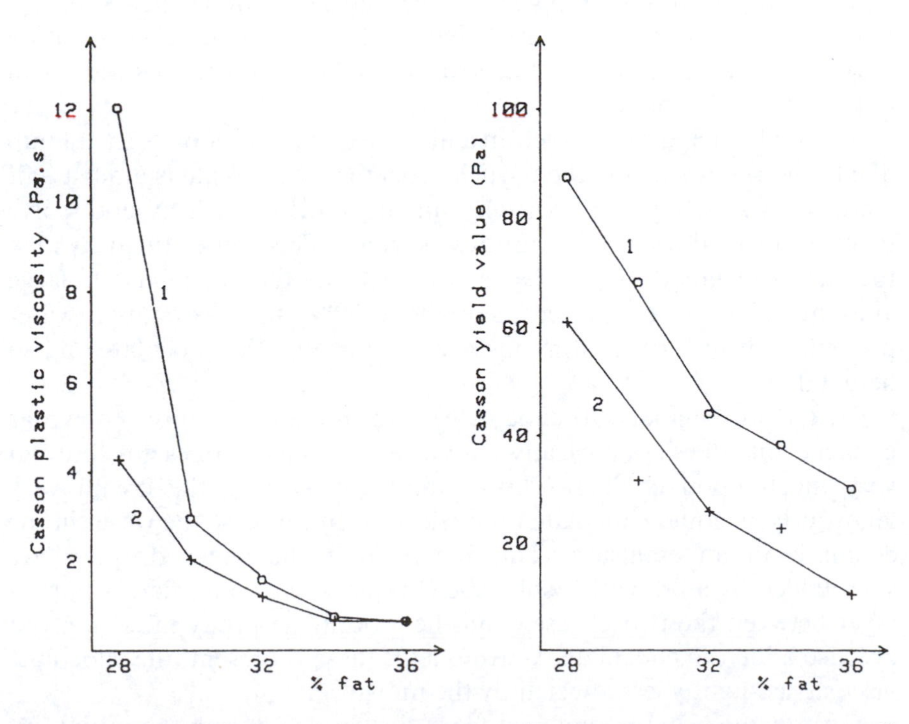

Figure 5.10 *The influence of fat content upon the viscosity parameters in two milk chocolates:* (1) *fine chocolate;* (2) *more coarse chocolate* (Chevalley[2])

partially tied to the particle surfaces. This free fat has a large effect on lubricating the flow when it takes place and so the plastic viscosity decreases dramatically. The yield value is more connected with the forces between the solid particles, which in turn are connected with the absolute distance between them. This will be less affected by the fat additions.

MOISTURE AND CHOCOLATE FLOW

It would be expected that, as water is a liquid, if it were added to liquid chocolate the resultant mixture would have a viscosity somewhere between those of the water and the chocolate. This is far from the case and the addition of 3% or 4% by weight of water will turn chocolate into a very thick paste (Project 8 in Chapter 10). Very approximately, for every 0.3% of extra moisture that is left within the chocolate at the end of conching, the manufacturer must add an extra 1% of fat. Because the fat is by far the most expensive major component within the chocolate, it is important that as much 'free' water is removed as possible.

Water is like the fat in that it can be bound or free, but unlike fat it should be as much as possible in the bound condition. If the total moisture is measured, for instance by Karl Fischer titration (see Chapter 8), then part of this is due to water of crystallisation in the lactose as it is often present as a monohydrate (Chapter 2). Other water may be inside the cocoa cells which have not been destroyed by milling. This water will not effect the chocolate flow.

If a bowl of icing sugar is left in a moist room it will soon form a lump due to the water sticking the particles together. Chocolate is almost half made up of very tiny sugar particles and any moisture will either dissolve them or form sticky patches on their surface. This causes them to stick together and greatly increases the viscosity of the chocolate. If large amounts of water are present, say above 20%, there is enough water present for it to form continuous streams through the chocolate and so help it flow.

Cream can be added to chocolate to form a soft mixture, known as ganache, that does not have any snap when broken and does not contract very much upon cooling. This is made by stirring the cream very vigorously as liquid chocolate is added to it. In this case the water in the cream is in an emulsion. This means that the water droplets are surrounded by a fat, with a substance known as an emulsifier forming a layer between the two. These emulsifiers occur naturally in cream, but are also added to chocolate. When one of these is present, the chocolate viscosity is slightly less affected by the moisture.

Most of the water is removed from the conche during the initial dry conching stage (see Chapter 4). This must be done with care, however, because if the water vapour comes out of the chocolate ingredients faster than it escapes from the conche, then it can condense to form droplets. These droplets can return into the chocolate and dissolve some of the sugar particles. These will stick together to form hard gritty lumps, once the moisture is eventually removed. This means that even if a chocolate has been correctly milled it will still taste very sandy in the final product.

EMULSIFIERS AND CHOCOLATE VISCOSITY

The role of an emulsifier is to form a barrier between two non-mixable substances. They play an important role in separating water globules in fat, in for instance in margarine, which has water droplets in fat (a water in oil emulsion), or in cream, which is fat droplets in water (an oil in water emulsion). As was shown in Chapter 2, the fat within cocoa beans can exist in both types of emulsions. In chocolate there is almost no water, so the emulsifier is somewhat different. Here we have sugar particles, which are hydrophilic but lipophobic, in other words, which attract water but tend to repel fat.

Liquid chocolate flows because the sugar and other solid particles are able to move past one another, so, as was described for the conching process, the surfaces have to be coated with fat. This is something that does not occur very naturally, so like the water emulsions, a substance that forms a layer between the two will greatly help the process. In this case the emulsifier coats the solid surface and forms a boundary layer between the two and is really a surface active agent rather than an emulsifier.

The mechanism by which a surface active agent works is illustrated in Figure 5.11. The individual molecules have a lipophilic (fat-liking) tail, which sticks out into the fat, where it wants to be. The other end is lipophobic and does not like the fat. It therefore keeps itself away from the fat as far as possible by attaching itself to the (also lipophobic) surface of the sugar. This can be compared with seaweed on a rock. There is a 'head' which attaches itself to a rock, and a long 'tail' which goes out into the sea. The 'tail' waves around and alters the flow around the rock. Just as there are many types of seaweed, there are also many types of surface active agents. Some have very large 'heads' which bind very strongly to the sugar, whereas others are less strongly attached and may even be removed by the addition of a different surface active agent. Similarly there are different lengths of 'tails', which affect the flow properties in different ways. This means that a surface active agent that is especially beneficial with regard to the yield value may be poor with respect to the plastic viscosity, and *vice versa*.

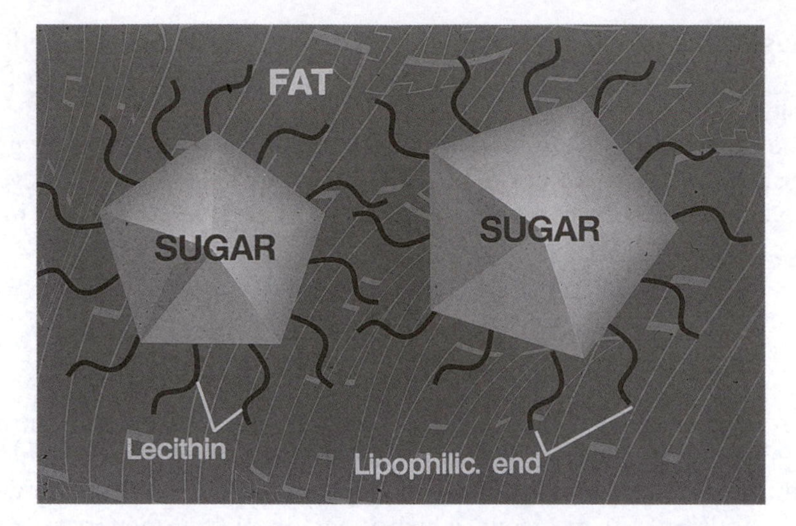

Figure 5.11 *Schematic diagram of lecithin molecules around a sugar particle*

Lecithin

The most common surface active agent is lecithin, which has been used in chocolate since the 1930s. It is a naturally occurring substance frequently obtained from soya and regarded by many as being beneficial to health. As described above it is able to attach itself to the sugar, whilst leaving the other end of the molecule free in the fat system to aid the flow. Harris showed that the lecithin was able to bind itself particularly strongly to the sugar, and it is this phenomenon which makes it so effective in chocolate manufacture. This was later confirmed by Vernier[3] (Figure 5.12) who used confocal laser scanning microscopy, which showed the fluorescing lecithin molecules to surround the sugar particles.

Additions of between 0.1% and 0.3% soya lecithin are said to reduce the viscosity by more than 10 times their own weight of cocoa butter. Also chocolates containing surface-active agents, such as lecithin, can tolerate higher levels of moisture than emulsifier-free ones. This is important because water is so very detrimental to chocolate viscosity.

Too much lecithin, however, can be detrimental to flow properties in that at higher levels, *e.g.* above 0.5% (Figure 5.13) the yield value increases with increased lecithin additions, although usually the plastic viscosity continues to fall. Bartusch[4] showed that at 0.5% about 85% of the sugar was already coated. After this the lecithin may be free to attach itself to itself to form micelles, or form a bi-layer around the sugar (so that the 'tails' of one layer of lecithin are separated by the 'tails' of a second layer facing the other way, as shown in Figure 5.14), either of which will hinder flow. The actual amount of lecithin that can be used

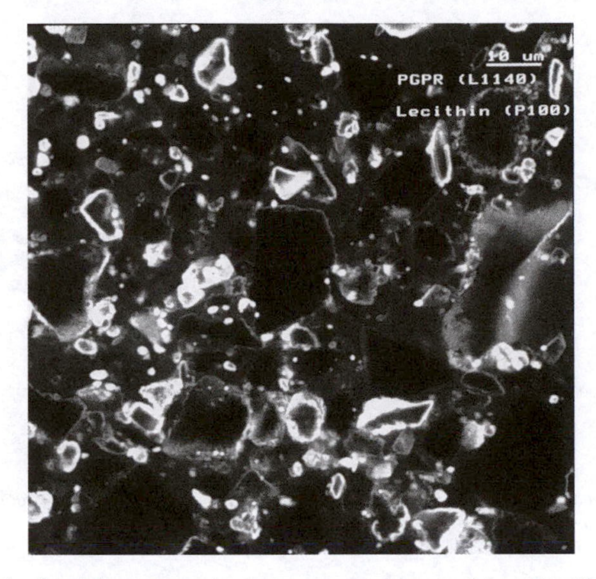

Figure 5.12 *Confocal laser scanning microscope picture of lecithin (fluorescing) surrounding solid particles within chocolate*

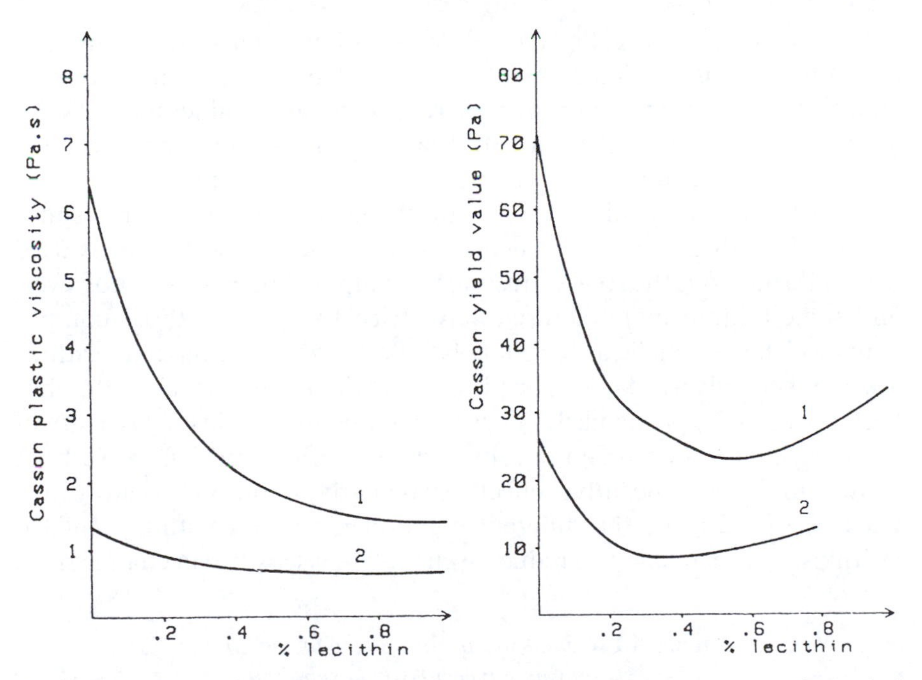

Figure 5.13 *The influence of soya lecithin upon the viscosity parameters of two dark chocolates:* (1) *33.5% fat;* (2) *39.5% fat* (Chevalley[2])

before thickening occurs depends to a certain extent upon the particle size distribution. A finely milled chocolate, with a large specific surface area, will have a relatively high yield value, as was explained earlier. This can be partially offset, however, by the fact that, as there is a bigger area

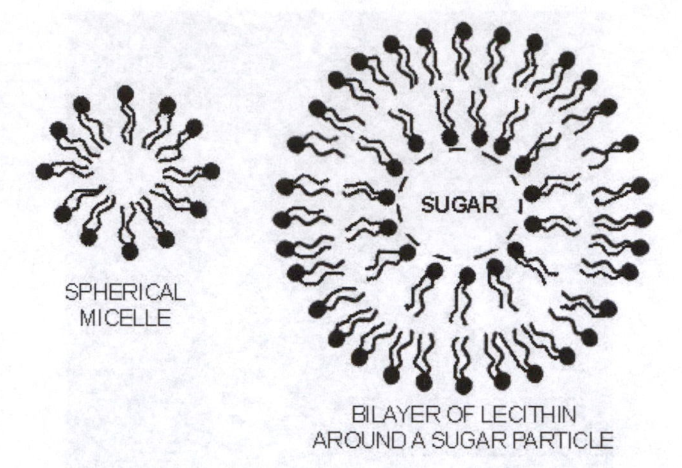

Figure 5.14 *Schematic diagram of a spherical micelle and a bi-layer of lecithin on a sugar particle*

to cover, more lecithin can be used before this increase in yield value takes place.

For a product to be called chocolate on its label in a shop the amount of lecithin used is restricted to 0.5% or 1.0% depending upon the type of chocolate that is being made and where it is being manufactured or sold. There is also a very small amount of lecithin which is present naturally in the cocoa and the milk components, especially in buttermilk.

Soya lecithin is a mixture of natural phosphoglycerides (phospholipids) with other substances such as soya oil (see Table 5.1). It is used widely throughout the food industry. Its composition can vary, however, and some lecithin manufacturers have tried to optimise those components which are beneficial for chocolate flow; thus fractionated lecithins are commercially available. The phosphatidylcholine part of lecithin has been shown to be particularly effective in reducing the plastic viscosity of some dark chocolates (Figure 5.15), whereas other fractions have been shown to have a negative effect particularly upon the yield value. Because the ratio of the different components varies within standard lecithins, its effectiveness in reducing chocolate viscosity can change from

Table 5.1 *Composition of soybean lecithin: phosphoglycerides (phospholipids)*

Phosphatidylcholine (PC)	13–16%
Phosphatidylethanolamine (PE)	14–17%
Phosphatidylinositol (PI)	11–14%
Phosphatidic acid (PA)	3–8%
Other phosphoglycerides	5–10%
The remaining approx. 44% is mainly triacylglycerols	

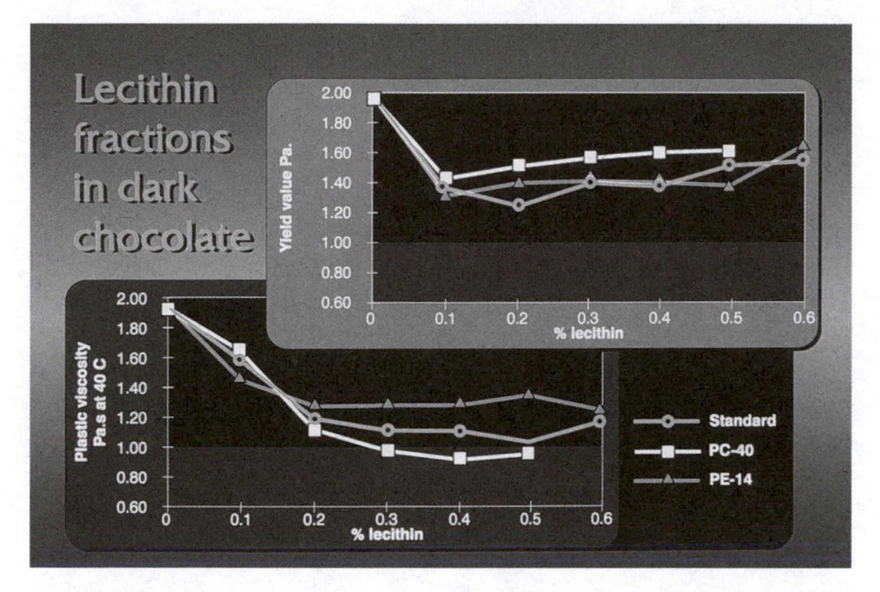

Figure 5.15 *The effect of fractionated lecithin upon the viscosity parameters of a milk chocolate*

batch to batch. For this reason some suppliers provide a standardised product.

In order to try to overcome the increase in yield value at higher levels, Cadbury developed an alternative surface active agent from hardened rapeseed oil. This is an ammonium phosphatide and is known as YN.

Commercial lecithin is primarily manufactured from soya and, although lecithin is generally regarded as being beneficial to health, some countries have expressed concern that part of the soya may be derived from a genetically modified source. This has led to lecithin being sold as being from other sources or selected growing areas such as Brazil. In addition alternatives to lecithin are being marketed for confectionery use, *e.g.* citric acid esters. Their actual effectiveness is likely to depend upon the chocolate type and the method of manufacture used to produce it.

Polyglycerol Polyricinoleate

Polyglycerol polricinoleate (PGPR), also known as Admul-WOL, is a very different surface active agent/emulsifier. It was originally developed for use in the baking industry and can be manufactured by the polycondensation of castor oil and of glycerol, the products of which are mixed and then esterified. Although having a relatively small effect on the plastic viscosity, it has a dramatic one upon the yield value. At only 0.2% in the chocolate it can halve the yield value relative to lecithin and at about 0.8% it has been shown to remove it altogether, turning the

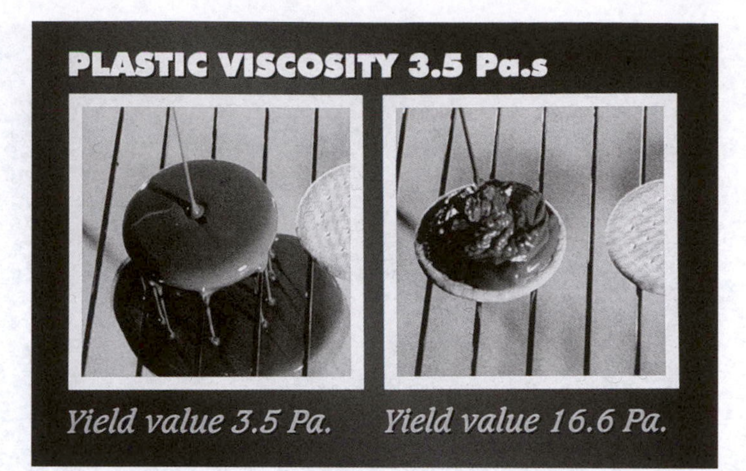

Figure 5.16 *Pictures illustrating the effect of the yield value when coating a biscuit with chocolate*
(Reproduced with the permission of Palsgaard, Denmark)

chocolate into a Newtonian liquid. How it does this has been the result of a lot of studies, but is still not fully understood. Its effect is very significant and may be useful or undesirable. Figure 5.16 shows two chocolates pouring on to a biscuit, with the same plastic viscosity, but very different yield values due to the presence in one of some PGPR. Very often the required flow properties lie somewhere between these two extremes, and mixtures of PGPR and lecithin are commonly used by many chocolate manufacturers and indeed sold as such by commercial emulsifier suppliers.

Other Emulsifiers

Other emulsifiers such as sorbitan esters, spans and tweens *etc.* are often found in chocolates and chocolate flavoured compound coating. They are normally less effective in reducing the yield value or the plastic viscosity than either lecithin or PGPR. They often do, however, alter the setting rate, the gloss on the product and especially its shelf-life with respect to bloom formation. This will be described in more detail in Chapter 6.

DEGREE OF MIXING

The degree of shear/mixing is critical in order to obtain the thinnest possible chocolate for a given fat content. This to a large extent is governed by the design of the conche and its mixing elements. There are other factors that are important, two of which are the order of addition

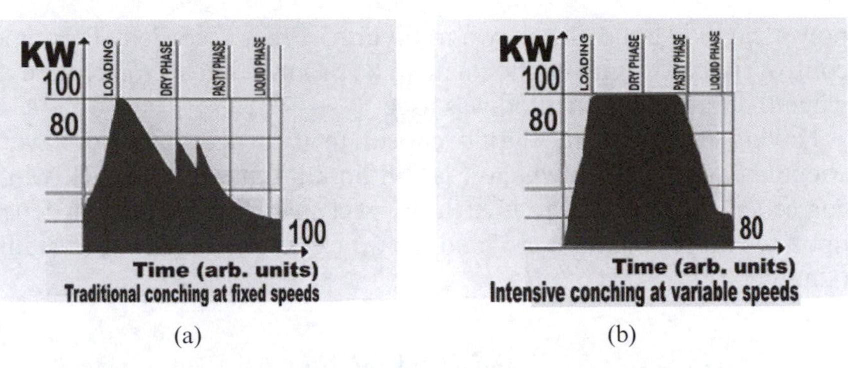

Figure 5.17 *Graph illustrating typical conche power amperage during the three stages of conching:* (a) *two speeds with reverse direction operation;* (b) *continuously adjusted speed*

of the ingredients into the conche and the drive controls to these mixing elements.

The conche is trying to coat the solid particles with fat and a smearing action does this very well, provided the particles are unable to move out of the way. If there is a lot of fat present they can move more easily and so the process becomes less efficient. This means that it is only necessary to add enough fat for the chocolate to turn into a paste. The remainder of the fat should be added towards the end in the liquid conching stage. It is said that fat added at this stage is twice as effective in reducing the final chocolate viscosity as the same amount of fat added at the beginning of conching.

It is even more important to add the majority of the lecithin at the later stages. One end of the lecithin molecule is very hygroscopic and will bind in water, making it more difficult to remove. This means that it should be added at the end of the dry conching stage, when a majority of the water has been removed, and preferably, like the fat, it should be added at the liquid conching stage. In addition some people believe that if lecithin is present at the roll refining stage, then the pressure will force it into the cocoa particles making it ineffective. Other sources state that high temperatures will reduce the effectiveness of lecithin, but this also has never been fully substantiated.

How the mixing elements are controlled will also effect the viscosity of the final liquid chocolate. Traditionally the mixing arms had one or two speeds, and could be reversed once the chocolate became thin, so that the wedge end cut into it (as described in Chapter 4). If the power used by the conche is recorded against time, this gives a jagged curve similar to the one illustrated in Figure 5.17a. For long periods therefore the conche is using very little power and therefore having very little mixing effect. Now it is possible to electronically control the process so that when the power drops, the speed is increased to bring it back up again. This means that a power curve like the one shown in Figure 5.17b is recorded. This type of control enables a thinner chocolate to be produced in a shorter time than when the traditional method was used.

Having produced the liquid chocolate it is necessary to have the chocolate in a form in which it is still liquid, but will set quickly in the correct crystalline form to give the correct snap and gloss. This depends upon the fat type (Chapter 6) and the way it is cooled and precrystallised (Chapter 7).

REFERENCES

1. R.B. Nelson and S.T. Beckett, 'Bulk Chocolate Handling', in S.T. Beckett (ed.), 'Industrial Chocolate Manufacture and Use', 3rd Edition, Blackwell, Oxford, UK, 1999.
2. J. Chevalley, 'Chocolate Flow Properties', in S.T. Beckett (ed.), 'Industrial Chocolate Manufacture and Use', 3rd Edition, Blackwell, Oxford, UK, 1999.
3. F. Vernier, 'Influence of Emulsifiers on the Rheology of Chocolate and Suspensions of Cocoa or Sugar Particles in Oil', PhD Thesis, Reading University, 1997.
4. W. Bartusch, 'First International Congress on Cacao and Chocolate Research', Munich, 1974, pp. 153–162.
5. S.T. Beckett, 'Conching', in S.T. Beckett (ed.), 'Industrial Chocolate Manufacture and Use', 3rd Edition, Blackwell, Oxford, UK, 1999.

Chapter 6

Crystallising the Fat in Chocolate

In order for a product to be sold as chocolate most of the fat inside it must be cocoa butter. This is a fat made up of several different triacylglycerols (*triglycerides*), each of which will solidify at a different temperature and at a different rate. To make it even more complicated there are six different ways that the individual crystals can pack together. What makes it difficult for the chocolate maker is that only one of these six forms will give the product the good gloss and snap upon breaking that makes it so attractive to the purchaser.

Milk chocolate must also contain milk fat. This will alter how it sets and also the texture of the final product. In some countries chocolate may contain a non-cocoa vegetable fat. When two or three fats are mixed together, the setting properties and texture are not a simple average of its components, as a phenomenon known as fat eutectics takes place. This means that there is a limit to the number of types of vegetable fats that can be used in chocolate.

If the wrong type of fat is present, or if the chocolate is old or has not been crystallised properly, a white powdery surface forms, known as fat bloom. This is in fact made up of fat crystals and not mould. Special fats and emulsifiers have been developed to retard its formation and to make chocolate more able to withstand higher temperatures. Chocolates that have been left in the sun will bloom very rapidly.

Some confectionery products contain a chocolate flavoured or compound coating. There are two main types of these, one which contains some cocoa butter/liquor and another which only has cocoa powder.

THE STRUCTURE OF COCOA BUTTER

All fats are mixtures of triglycerides: that is they have three fatty acids attached to a glycerol backbone. In cocoa butter there are three main acids which account for over 95% of those present. Almost 35% is oleic acid (C18:1), about 34% is stearic acid (C18:0) and approximately 26%

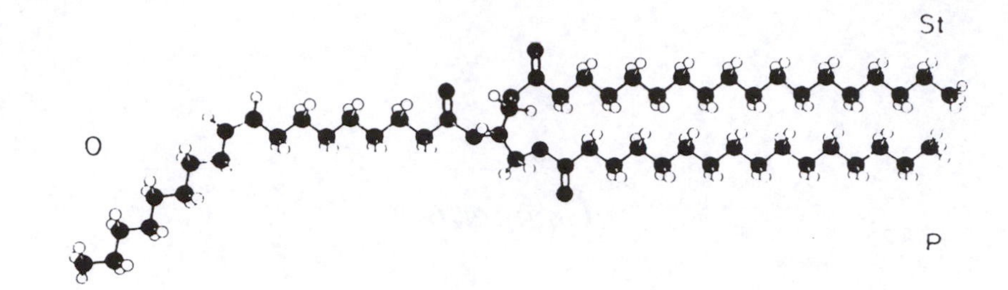

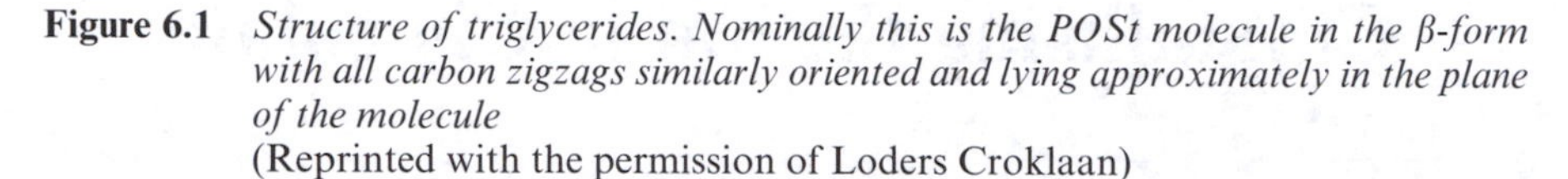

Figure 6.1 *Structure of triglycerides. Nominally this is the POSt molecule in the β-form with all carbon zigzags similarly oriented and lying approximately in the plane of the molecule*
(Reprinted with the permission of Loders Croklaan)

is palmitic acid (C16:0). It is in fact because this fat is relatively simple in having so few main components that it melts rapidly over such a small temperature range, *i.e.* between room and mouth temperatures.

These acids are attached to the glycerol in a way that is illustrated diagrammatically in Figure 6.1. This shows palmitic acid (P) in position 1, oleic acid (O) in the 2 position and stearic (St) in position 3. This is known as a POSt molecule. If the stearic and oleic acids were reversed this would become PStO, which is quite a different molecule even though the constituents are the same.

The stearic and palmitic acids are saturated acids, that is the hydrocarbon chain which makes up the fat does not contain any double bonds. In unsaturated fats this chain contains one or more double bonds, as is the case for oleic acid. The molecule in Figure 6.1 can therefore be described as symmetrical mono-unsaturated and is often referred to as an SOS triglyceride, where S refers to any saturated acid. About 80% of cocoa butter is of this form, *i.e.* it has oleic as the middle acid.

Between about 1% and 2% of cocoa butter is all saturated (SSS, long chain trisaturated triglycerides, where the saturated fat is mainly palmitic or stearic) and melts at much higher temperature than the more common SOS. From 5% to almost 20% on the other hand contain two oleic acid molecules and are SOO, which is mainly liquid at room temperature. When these are combined, as they are in cocoa butter, the fat will therefore be partly liquid at room temperature. If milk fat is present this proportion will increase and the chocolate will be softer to bite into. As the temperature rises the fat will melt according to the proportions of the different types of fat present.

It is possible to measure the proportion of solid fat present at any temperature using techniques such as nuclear magnetic resonance (NMR, see Chapter 8) and the curve produced is known as the solid fat content. Figure 6.2 shows examples of this for cocoa butters from three different countries: Brazil, Ghana and Malaysia. In Chapter 2 it

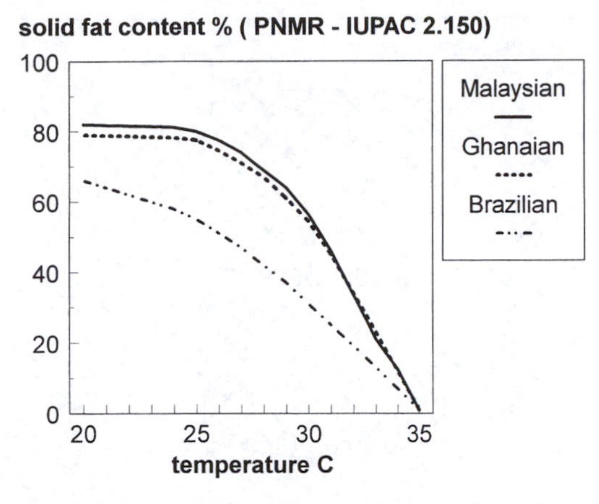

Figure 6.2 *Typical solid fat content of cocoa butters from Brazil, West Africa and Malaysia obtained by NMR measurements*

was stated that normally the nearer the equator the cocoa was grown the harder will be the fat. This is demonstrated clearly here, as at 20 °C 81% of the Malaysian cocoa butter is solid compared with only 66% for the Brazilian one, with the Ghanaian one being in between. At 32.5 °C the difference is proportionately greater with only 7% of the Brazilian sample still being solid compared with 20% for the Malaysian one. It is possible to see the reason for this by looking at the triglyceride content of the different cocoa butters in terms of their saturated and unsaturated fats, as is given in Table 6.1. The relative hardness is particularly affected by the SOS/SOO ratio. This is 3.6 for Brazil but 16.5 for Malaysia.

When the cocoa butter is melted and then cooled the three types of triglyceride once again behave differently. This is illustrated in Figure 6.3. The SSS types crystallise first. These make the chocolate thicker as there is less liquid fat present, but it is the SOS crystals that form

Table 6.1 *Triglyceride composition (%) of cocoa butters from different growing areas*
(Reprinted with the permission of Loders Croklaan)

Triglyceride	*Brazil*	*Ghana*	*Malaysia*
SSS	1.0	1.4	2.3
SOS	63.7	76.8	84.0
SSO	0.5	0.4	0.5
SLS	8.9	6.9	6.8
SOO	17.9	8.4	5.1
OOO	8.0	6.1	1.3

S = Saturated fatty acids (mainly palmitic and stearic); O = oleic acid; L = linoleic acid.

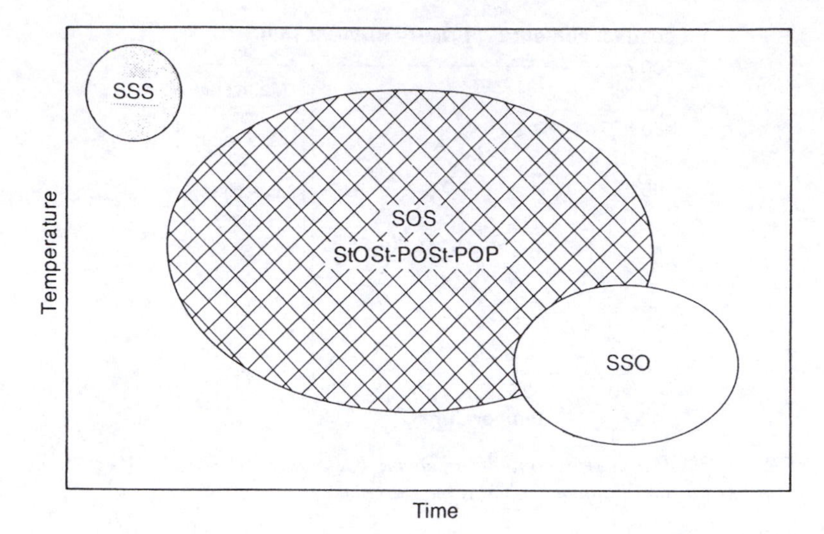

Figure 6.3 *Crystallisation temperatures and rates for the different triglyceride groups* (Talbot[1])

later on that determine the chocolate's texture and resistance to fat bloom.

DIFFERENT CRYSTALLINE FORMS

It is commonly known that carbon can exist in different forms, ranging from soft graphite through intermediate hardness materials up to the very hard structure of diamonds. Fats can also crystallise in a number of different ways, a property that is known as polymorphism. As the structure becomes denser and lower in energy, it becomes more stable and harder to melt.

The reason for this is that the different fat molecules, as were illustrated in Figure 6.1, can fit together in a number of different ways. Because of their shape it is in a way like stacking chairs. There are two

Double-chain length packing Triple-chain length packing

Figure 6.4 *Double- and triple-chain length packing configuration* (Reprinted with permission of Loders Croklaan)

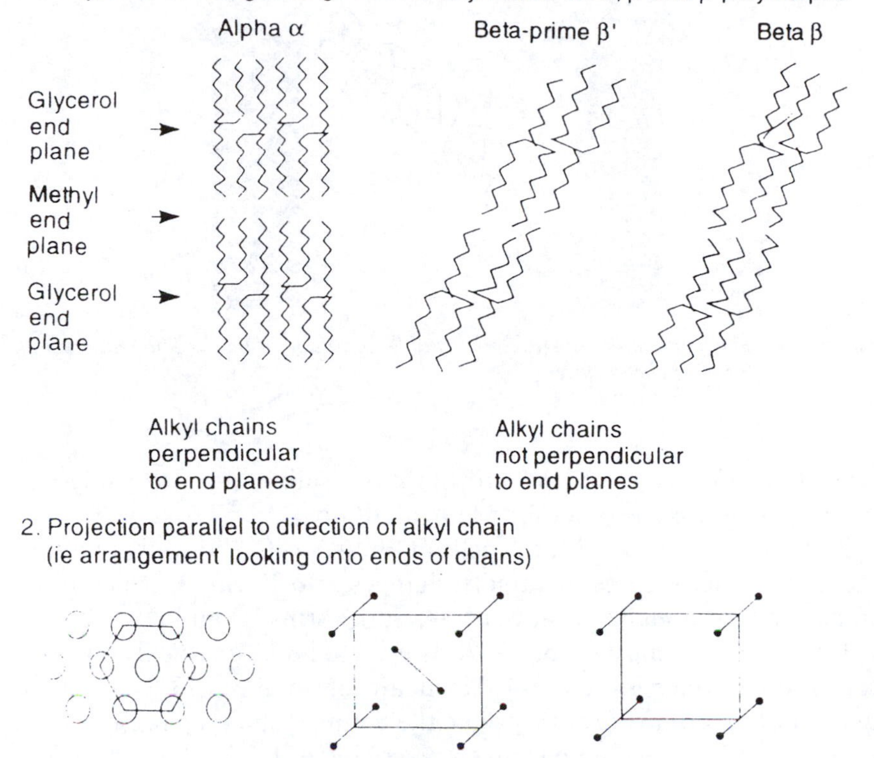

Figure 6.5 *Crystal packing of tryglycerides.* (1) *Projection showing arrangement of alkyl chains for α, β and β′ polymorphs;* (2) *projection looking onto the ends of chains* (Reprinted with the permission of Loders Croklaan)

ways that this can be done, as is illustrated in Figure 6.4, that is double chain packing and triple chain packing. These small stacks have then got to fit together with other stacks. The angle at which they fit together determines their stability. If chairs are stacked straight up they tend to fall over. This is the same for fats in that a straight up arrangement, known as α, is formed at low temperatures and transforms rapidly into one of the other forms (see Figure 6.5).

The analogy with chairs then breaks down. The angle of tilt at which the crystals fit together then determines their stability, whereas chairs stacked at an angle would just fall over. Some fats have only one stable form whereas others have three, namely α, *β′* and *β*. Cocoa butter has six, however. There are two nomenclatures that are used to describe these polymorphs. The chocolate industry tends to number them I to VI, as was described by Wille and Lutton in 1966. The oils and fats industry in general prefers the Greek letters defined by Larsson in the same year. Figure 6.6 shows the temperature ranges at which the different forms

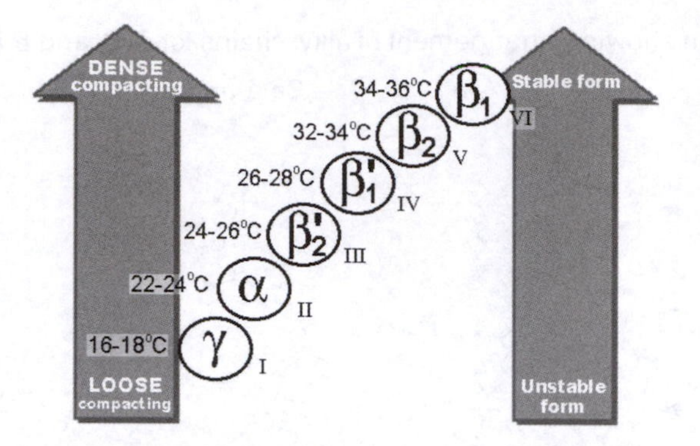

Figure 6.6 *The temperature ranges for the stable formation of the six different crystalline forms of cocoa butter*

crystallise. Forms V and VI are the most stable and are triple chain packing, whereas the other forms are all double.

Form 1 is very unstable and melts at about 17 °C, so is only present on coatings for ice creams. It rapidly changes into Form II, which in turn transforms, although at a slower rate, into Forms III and IV.

If liquid chocolate at about 30 °C is used to make a typical confectionery product, which is then cooled in an airflow at about 13 °C for about a quarter of an hour, Form IV will be the main crystal type present, unless some sort of pre-seeding has taken place. Form IV is relatively soft and so the chocolate will not have any snap when it is broken. In addition it will transform over a period of days to Form V. The actual time depends on the storage conditions, with the transformation taking place faster at higher temperatures. The more stable forms are, however, more dense, so the chocolate will contract. Some of the cocoa butter is still liquid even at room temperature, however, and in addition to this some energy is given out as the fat is transforming to a lower energy state. This combination of effects pushes some of the fat between the solid particles and on to the surface. Here it forms large crystals that give the white appearance of chocolate bloom, as is shown in Figure 6.7.

For this reason it is necessary for the chocolate maker to ensure that the cocoa butter is in Form V when it is used to make confectionery products. This form is hard with good snap and gives a glossy appearance with a relatively good resistance to bloom. It will also contract well if the liquid chocolate is poured into a mould (see Chapter 7).

Form VI is in fact more stable, but under normal conditions is only formed by a solid to solid transformation and not directly from liquid cocoa butter. This means that chocolate with fat in Form V will after a period of months, or sometimes years, start to bloom. This is because the same effects as were described for the IV to V transition are happening again, but at a slower rate. Therefore the skill of the chocolate maker is

Figure 6.7 *Picture of bloomed and unbloomed dark chocolate sweets*

to get the chocolate as rapidly as possible into form V, but then take precautions to prevent the further transformation.

PRECRYSTALLISATION OR TEMPERING

If chocolate is cooled to 34 °C and then stirred slowly, Form V crystals will eventually appear and after a long time, probably days, there will be enough to seed all the remaining chocolate. This means that if it is cooled suddenly there are enough crystals present to form nuclei around which the remaining fat will set with the same type of crystal. This is obviously impractical for the chocolate industry, where several tons of chocolate an hour are often used.

Some other form of pre-crystallising, commonly known as tempering is required. For small batches it is possible to add small amounts of previously set chocolate. Here a few percent of grated solid chocolate is added to liquid chocolate which has been previously cooled to about 30 °C. This is often described in cooking recipes for the home, when cocoa butter-containing chocolates are being used (see Project 9 in Chapter 10). (Many cooking chocolates are in fact compounds which contain other fats. These fats only solidify in one crystalline form so seeding or tempering is not required.) Recently a method has been developed to produce small cocoa butter crystals by spray chilling. Once they have transformed to Form VI, they are used to seed chocolate.

The speed with which the fat in chocolate begins to crystallise not only depends upon the temperature but also upon the rate at which it is mixed and sheared. This is due to the fact that fat will solidify on to any seed crystal that is present. The seed therefore needs to be of the correct type and to be well distributed throughout the chocolate. Large crystals will have much less effect than the same amount of solid fat in small crystals that are uniformly mixed throughout the chocolate. High shear has the effect of breaking the solid fat crystals and uniformly distributing them. In addition it provides heat and energy which increase the rate in which the more unstable crystals can change to Form V. Ziegleder showed that the effect was extremely dramatic and some of his results are given in Figure 6.8. This shows that by using extremely high shear rates cocoa butter can be precrystallised in 30 s rather than several hours or days.

There is a major problem with high shear rates, however, in that they can generate too much heat, which will help the transformation to Form V. If there is too much heat, on the other hand, it will melt all the crystals altogether.

The actual process used by chocolate manufacturers is a compromise that uses an intermediate shear rate. In order to increase the rate of crystallisation the chocolate is cooled to temperatures where Forms II and III are created. It is thoroughly sheared to form a lot of smaller crystals, before being reheated, still under shear, to convert the unstable

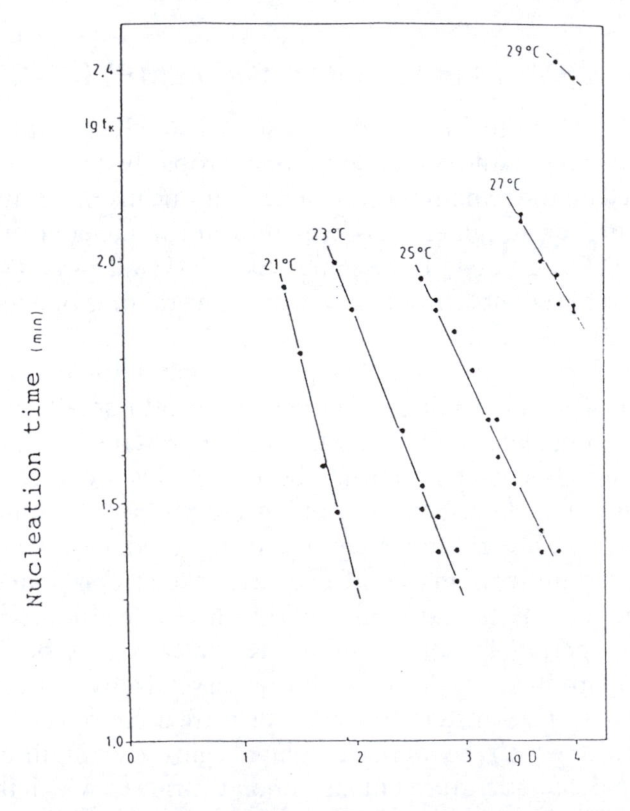

Figure 6.8 *The relationship between crystallisation and the rate of shear (mixing) at different temperatures* (Ziegleder[2])

ones to the required form. The machines that do this are called tempering machines and further details about them are given in the next chapter. The actual temperatures used depend very strongly upon the fats present in the chocolate. These are mainly cocoa butter and milk fat, but in some chocolates non-cocoa vegetable fats are also present.

MIXING DIFFERENT FATS (FAT EUTECTICS)

When two or more fats are mixed together it is important that the final chocolate sets at a suitable rate and, more importantly, that it has the correct texture and melting properties in the mouth. One way to determine this is to measure the solid fat content, as was demonstrated for different cocoa butters in Figure 6.2. If instead of measuring the proportion of solid fat at different temperatures the proportion is determined for different blends of two fats at the same temperature then a plot is obtained as is illustrated in Figure 6.9.

It might be expected that the solid fat content could be calculated from the solid fat measurement for the individual fats according to the proportion in which they are present in the mixture, *i.e.* a straight line on the graph. As can be seen this is not the case.

The reason for this is that although other fats, like milk fat, are triglycerides, their structure is very different from that of cocoa butter.

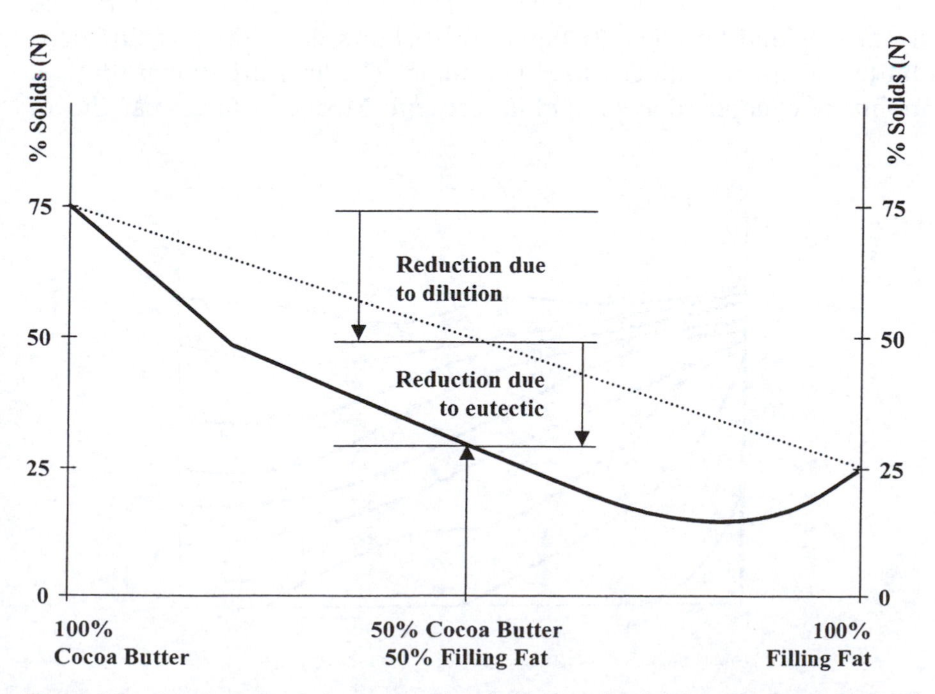

Figure 6.9 *Graph showing the expected and the actual solid fat content at 20 °C of mixtures of cocoa butter and a soft filling fat*

Returning to the analogy of stacking chairs, it is as if chairs of another size are placed within the stack. This will make the overall structure far less stable. This means the product will melt more easily, that is it will contain far more liquid fat, and this is what in fact happens for most fat mixtures, as is shown in Figure 6.9.

If only a small amount of the other fat is present the disruption it causes is less, so the actual hardness is near the expected one. Where the two fats are in a similar proportion the softening effect is largest. The degree of this difference and also the amount that can be added without causing a significant texture difference will depend on how differently the fats crystallise. As will be seen later this is very important when adding vegetable fats, and very few can in fact be used.

Milk fat is present in all milk chocolates and even many plain ones. In the second case the reason is to reduce the possibility of fat bloom. If milk fat is added at about 5% of the weight of the chocolate, which contains a total of 30% fat, then it in fact makes up about 17% of the fat phase. This makes it softer and also increases the time that it takes for the cocoa butter to transform from Form V to Form VI and produce a white coating on the surface.

In milk chocolate even higher levels of milk fat are sometimes used, but there is a limit, which is set by the softening (eutectic) effect. Chocolate eaters expect milk chocolate to be softer than dark chocolate, but they don't want it to be too soft. This once again can be determined from the solid fat content. Figure 6.10 shows data for milk fat/cocoa butter mixtures at different temperatures. The lines are drawn through points of equal hardness/liquid fat content. Most consumers eat choco-

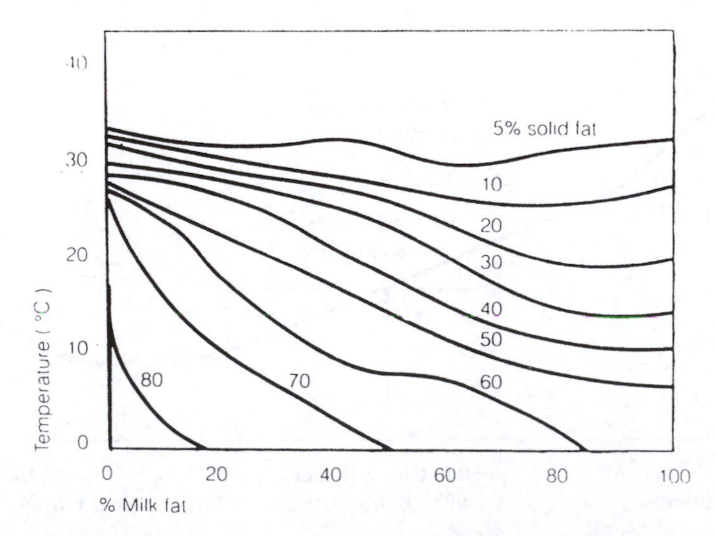

Figure 6.10 *Melting profiles of fat blends of cocoa butter and milk fat* (Reprinted with the permission of Loders Croklaan)

late at around 20 °C, and for it to be acceptable it must have about 70% solid fat. This means that about 15% milk is an upper limit. This is of course for fat that is freely mixed with the cocoa butter. In looking at the different types of milk powders in Chapter 2 it was noted that some milk fat can be retained within the structure of the powder. This was said to be detrimental to the flow properties of the liquid chocolate, which it is. On the other hand any bound fat does not contribute to the softening effects. Chocolates containing some bound fat have been manufactured with the milk fat accounting for over 27% of the fat phase.

The solid fat content not only tells us about the hardness of the chocolate, but also gives information about what happens when it melts in the mouth. Figure 6.11 illustrates curves for three different fat mixtures. The percentage solids at room temperature gives an indication of the hardness of the chocolate; in this case A is the hardest and C the softest. The temperature at which the sudden reduction in solid fat content occurs relates to its heat resistance. Malaysian cocoa butter would normally be at a higher temperature than Brazilian butter (Figure 6.2). This makes it better for use in hot climates, but worse for ice-cream. In this case once again A melts at a higher temperature.

The steepness of the curve relates to how quickly it will melt. To change from a solid to a liquid requires a lot of energy. This is known as latent heat and for cocoa butter is about 157 J g^{-1}. This can be compared with the specific heat of 2.0 J g^{-1}, which is the amount of energy required to raise the temperature of the fat by 1 °C. When

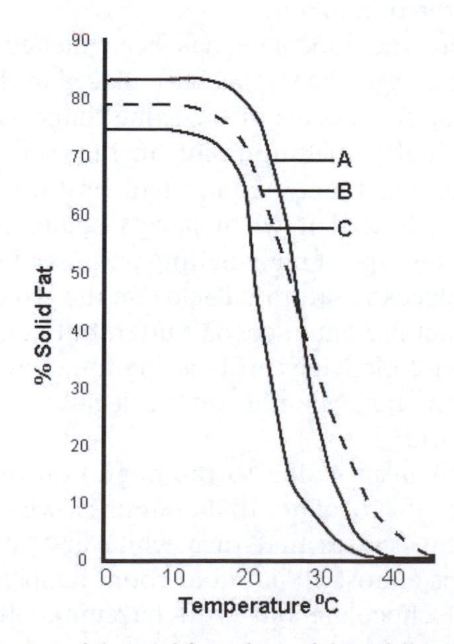

Figure 6.11 *An illustration of the solid fat indices of three different fat mixtures*

chocolate melts in the mouth, therefore, a lot of energy is required, which can only come from the mouth itself (The temperature is raised by approximately 20 °C and it also melts; the combined specific and latent heat amounts to: $20 \times 2 + 157 = 197\ \mathrm{J\,g^{-1}}$). If the slope is steep this happens very quickly and the mouth feels cool. Fats have been developed with exceptionally sharp melting curves, which exploit this property. The speed of melt also alters the viscosity of the chocolate before it is swallowed, which in turn changes the speed with which the particles can reach the flavour receptors. So the flavour is also changed. In the diagram A and C will have similar cooling effects, both of which are greater than that of B.

Some fats, like B, have a significant amount of solid crystals still present even at 40 °C. This will not melt in the mouth and will leave a waxiness in the mouth. Some fats used to make chocolate flavoured coatings suffer from this problem. Other fats can be added which will reduce the amount of these high melting point fats by the eutectic effect. Care must be taken, however, as this can lead to chocolate bloom formation.

CHOCOLATE FAT BLOOM

There are four main ways in which chocolate fat bloom is formed. Two have already been described, *i.e.* by the Form IV to Form V transformation following incorrect precrystallisation (tempering) or by the age- and temperature-related Form V to Form VI change, which can be slowed down by the addition of milk fat.

Another is where the chocolate has been melted and recrystallises without retempering, *e.g.* when it has been placed in the sun. This can be overcome by adding fat crystals of the same shape as cocoa butter, but with a very much higher melting point, to the chocolate. Unless these crystals are melted, which may require temperatures above 50 °C, they remain in it and will seed it when it sets again, preventing it from blooming. One of this type of high melting point seeding fat is made with behenic acid to replace the saturated acids on the glycerol chain. Behenic acid (C22:0) is in fact present in cocoa butter, but at much less than 1%. BOB (1,3-behenoyl-2-oleoylglycerol) is manufactured by the Fuji Oil Company in Japan, but can not yet be legally used in many other countries of the world.

The fourth mechanism is due to the migration of soft fats into the chocolate. In a box of chocolates that contains a wide variety of centres, it is usually the nut centres that turn white first. Nuts like hazelnuts contain a fat that is almost all liquid at room temperature. Pralines are made similarly to chocolate but with hazelnuts. If the two are put together to form a sweet, the fat phase tries to reach an equilibrium (see

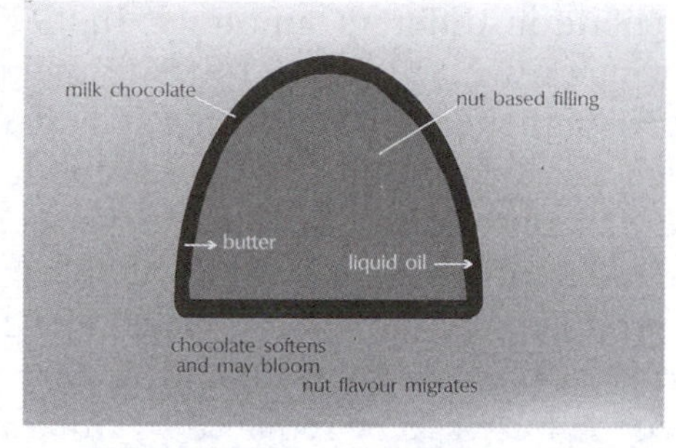

Figure 6.12 *Illustration of the migration of fats within a chocolate with a nut based filling*

Figure 6.12). The liquid fat from the centre moves into the chocolate and because of the eutectic effect makes it much softer and may even cause some of the cocoa butter to become liquid again, if the softening is bad enough. Some of the cocoa butter migrates in the reverse direction into the centre making that harder. Thus the difference in texture is lost and the sweet is less appealing. As the soft fat migrates to the surface it takes some of the cocoa butter with it. This crystallises forming fat bloom. It is also possible that this type of fat causes a rapid increase in the rate of the Form V to Form VI transition of the cocoa butter, which would also cause it to bloom.

There are several ways of reducing this type of bloom. Some are based on trying to stop the soft fat getting into the chocolate at all. This is done by having another harder fat that either forms a layer inside the chocolate shell or forms a sponge-like system within the centre. Alternatively an anti-bloom fat is added to the chocolate or to the centre, from where it migrates together with the soft nut oil. These fats have the property that they can slow down the Form V to Form VI transition, which would cause the fat bloom.

These are just some of the vegetable fats that can be used in confectionery. Others may be added to increase the heat resistance or to remove the need for tempering.

SOME TYPES OF NON-COCOA VEGETABLE FAT

Vegetable fats have been used in chocolate and chocolate-like coatings for many years. In the First World War, companies like Rowntree used vegetable fats in their chocolate because they were unable to buy cocoa butter. In the 1950s research showed that, unlike animal fats, some vegetable fats contained the same triglycerides as cocoa butter. This led to a 1956 Unilever patent, which showed a method of producing a fat that was almost identical to cocoa butter, but was made from fat obtained from other vegetable sources. These were made commercially and added to chocolate in different amounts. In 1977 legislation was passed in the UK which restricted its use to 5% if the product was to be sold as chocolate. If higher levels were used it had to be sold under another name, such as a chocolate flavoured coating. Several other countries also adopted this legislation, but some such as France and Germany, required different labelling if any vegetable fat at all is present, even though the products containing the vegetable fat can be indistinguishable in terms of processing, taste and texture. A few countries even allow all the cocoa butter to be replaced by other fats.

The original vegetable fat made by Unilever and the many others that are now on the market are known as cocoa butter equivalents (CBEs), as they are like cocoa butter and can be added in any proportion without causing a significant softening or hardening effect. Other fats can only be used if almost all the cocoa butter is replaced, and these are known as cocoa butter replacers (CBRs).

Cocoa Butter Equivalents

To be able to be added to cocoa butter without having a eutectic effect, the vegetable fat must crystallise in the same way as cocoa butter (*i.e.* have the same size and shape chair). Cocoa butter contains palmitic (P), stearic (S) and oleic (O) acids on a glycerol backbone, with the majority of the molecules being POP, POSt and StOSt. The fat manufacturer has therefore to obtain these different fractions from different sources and then blend them.

The POP is the easiest to find as it is a major component of palm oil, which is obtained from the palm (*Eleaeis guineensis*) widely grown in Malaysia. A lot of other fats are present, but these can be removed by fractionation. The easily melting (olein) and the hardest melting (stearin) parts are removed, leaving the mid-fraction, which is mainly POP and a small amount of POSt. Two types of fractionation are used: dry and solvent fractionation. In dry fractionation the fat is heated to a predefined temperature and then the liquid part separated from the solid by pressing or filtering. In solvent fractionation, the fat is dissolved normally in acetone or hexane. The higher melting triglycerides are then allowed to crystallise and are filtered out. This procedure gives much more clearly defined fractions than the dry process.

The StOSt and increased amounts of POSt are much harder to obtain. One nut that gives this type of fat is illipe (*Shorea stenoptra*). This is grown in Borneo, but is only intermittently available. Shea (*Butyrospermum parkii*), which grows in West Africa, and sal (*Shorea robusta*) from India both contain a high proportion of StOSt, but the crop is not always available and can be of poor quality. However, by blending the palm oil mid fraction with illipe fat and the harder (stearin) fraction from shea fat, it is possible to manufacture a fat that is totally compatible with cocoa butter.

In summer in Southern Europe or in tropical climates, the chocolate will melt easily under ambient conditions. By altering the proportions of StOSt it is possible to make the chocolate so that it will not melt until the temperature is several degrees higher than for normal cocoa butter, but does not leave a waxiness in the mouth. If the solid fat content were measured as in Figure 6.11, the nearly vertical melting curve would move to the right and there would be no significant amount that was solid above about 36 °C. This type of fat is known as a cocoa butter improver (CBI), because it improves the chocolate. The harder fat components are difficult to obtain, so CBIs are more expensive than cocoa butter equivalents.

Enzyme Interesterification

Because good quality harder fats are difficult to obtain, fat manufacturers have developed a process that enables them to use other raw materials such as sunflower oil. This is used to obtain a fat that can be blended with palm oil to give the required cocoa butter equivalent.

Enzymes occur naturally both inside and outside the human body and have the ability of being able to speed up the separation of the fatty acids from the glycerol backbone leaving partial glycerides (mono- and diglycerides) as well as glycerol. One of these, called lipase, occurs in flour and can give big problems to the confectionery manufacturer, as with certain fats it produces free fatty acids, which have a very unpleasant cheesy flavour. Some of these enzymes are position specific and will only accelerate that reaction which cuts off acids from specific positions on the glycerol. They are also able to exchange the fatty acids in certain positions as well as remove them. The fat manufacturers use this property to change the melting properties of their fats.

In the 1970s workers at Unilever Research Laboratory at Colworth House in England showed that a certain type of enzyme, called mucor miehei, only attacked the 1 and 3 positions of a triglyceride. The process they developed can use any fat that mainly has oleic acid in the 2 position (see Figure 6.13). This is then mixed with stearic acid and the appropriate

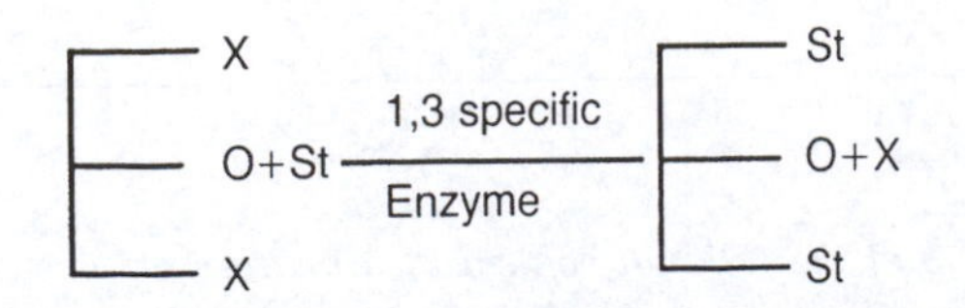

Figure 6.13 *Diagramatic representation of enzyme interesterification*

enzyme. This free stearic acid is then esterified on to the 1 and 3 positions, where the original fats are removed by the enzyme. The central oleic acid remains untouched, so the fat becomes very rich in StOSt. The free fatty acids released by the enzymes are then treated by de-acidification and then fractionation and refining to remove the impurities. This fat is identical to that obtained directly from other tropical nuts in terms of its processing to make confectionery and its behaviour in the mouth when eaten. It can also be added to chocolate at any level that is required, as it is fully compatible with cocoa butter. Some other fats are very different, however.

Lauric Fat Cocoa Butter Replacers

There are other fats that melt in the same temperature range as cocoa butter, and so have a similar texture and mouthfeel, but which crystallise in a very different way. Palm kernel oil and coconut oil are both widely available and contain about 50% lauric acid (C12:0, dodecanoic acid), a lot of which exists as trilaurin. Unlike cocoa butter it sets in one crystal form and so no pre-crystallisation is required. This form is also different in that it is not a β type crystal (like Form V in cocoa butter) but a β' one (equivalent to Form IV).

Once again we are in the situation of trying to stack different chairs. This is demonstrated clearly by measuring the solid fat content of mixtures of a lauric fat with cocoa butter as is shown in Figure 6.14. Apart from where only about 5% of the other fat is present, this mixed

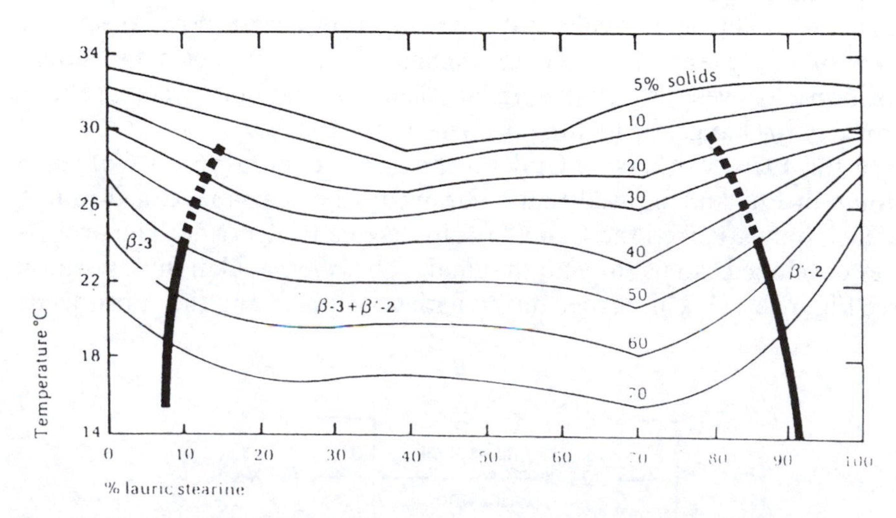

Figure 6.14 *Diagram of the solid fat content of a mixture of cocoa butter and a lauric CBR*
(Reprinted with the permission of Loders Croklaan)

crystal system exists. It is very soft, will take a long time to solidify and is likely to bloom at a very rapid rate.

This means that these lauric cocoa butter replacers can only be used when very little cocoa butter is present. Because cocoa liquor contains about 55% of cocoa butter, the products are normally made with cocoa powder. This in turn tends to give them a different flavour. They are, however, often made into coatings, because for the small confectioner or for home use it is a big advantage not to have to temper it. Cooling can be very rapid and the products are often initially very glossy.

With lauric fat products, however, it is very important to keep them in a dry environment and if possible use other ingredients that are lipase free. This is because in a moist environment the lipase enzymes will accelerate the removal of some of the free fatty acids from the glycerol backbone. In this particular case, the acids produced have a very unpleasant soapy flavour, even at a very low level in the product. Other fats do exist, however, which have slightly more compatibility with cocoa butter.

Non-Lauric Fat Cocoa Butter Replacers

Palm oil and soybean oil contain many of the same fatty acids as cocoa butter and these can be fractionated out leaving mainly the stearic, palmitic and oleic acids. These, however, are far more random than in cocoa butter, so for instance the oleic acid is very often in the 1 or 3 position. In addition they often contain a significant amount of elaidic acid (C18.1, octadec-*trans*-9-enoic acid). This is an unsaturated fatty acid, which is in the *trans* form. This means that the hydrogen atoms that are associated with the unsaturated carbon atoms lie on opposite sides of the double bond as illustrated in Figure 6.15 and results in its structure being very different from the unsaturated fat in cocoa butter, oleic acid (C18.1, octadec-*cis*–9-enoic acid). The *cis* form has the hydrogens on the same side of the bond as is also illustrated in Figure 6.15.

This more random structure coupled with this different unsaturated acid means that this type of fat has limited compatibility with cocoa butter. Measurements of the solid fat content for different mixtures are given in Figure 6.16. Once again the soft, likely to bloom conditions occur when both fats are present in similar amounts. However, the effect

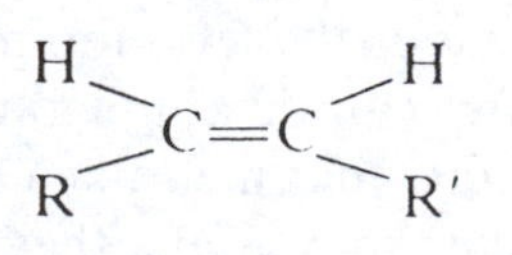

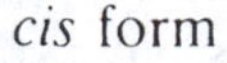

H R′
C=C
R H
trans form

Figure 6.15 *Illustration of the difference between* cis *and* trans *forms*

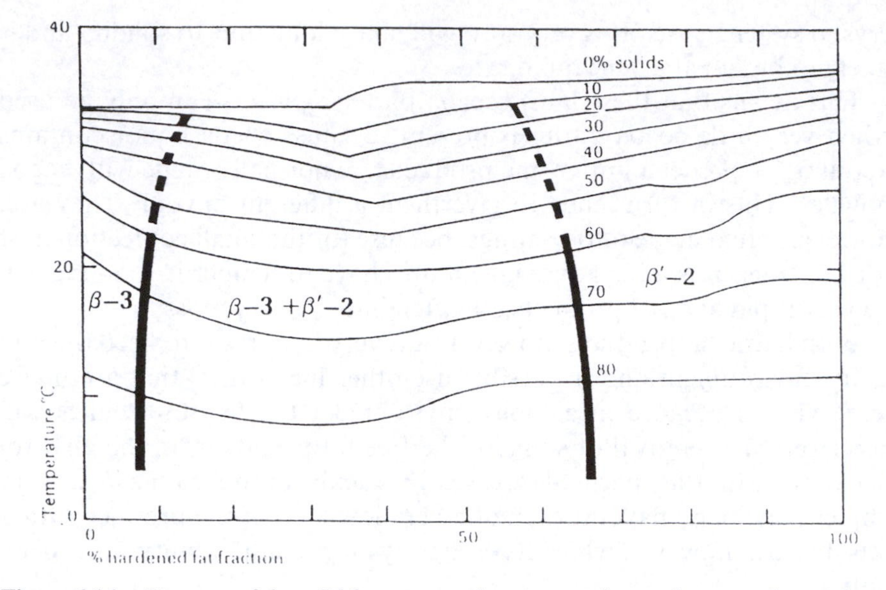

Figure 6.16 *Diagram of the solid fat content of a mixture of cocoa butter and a non-lauric CBR*
(Reprinted with the permission of Loders Croklaan)

is less severe than for lauric fats. In addition about 7% of lauric fat can be added to cocoa butter but, more importantly for the chocolate manufacturer, up to 25% of cocoa butter can be present with a non-lauric cocoa butter replacer. This means that chocolate flavoured coatings can be made with cocoa liquor, which in turn makes the flavour more like normal chocolate.

Non-lauric fats, like lauric ones, tend to set in the β' form and do not need tempering, although it may be better to do so if a significant amount of cocoa butter is present. They also need much slower cooling than the lauric fat cocoa butter replacers.

Low Calorie Fats

In the USA in particular, there has been a demand for low calorie products. Normal fat contains nine calories per gram compared with five calories per gram for the sugar and protein components. If the proportion of fat is reduced the calorific value will fall. For processing and texture reasons, however, it is not possible to reduce the level too much below 25%. This is insufficient to make a low calorie claim on the product, so two manufacturers have produced fats that melt like cocoa butter but have a lower calorific value. Like lauric fats they are incompatible with cocoa butter and so the products have to be made with cocoa powder.

Procter & Gamble produce a fat called caprenin from caprylic (C8), capric (C10) and behenic (C22) acids. Caprylic and capric acids are metabolised in the body in a different way from the cocoa butter fatty acids, whilst, because of its chain length, behenic acid is poorly absorbed at all. This means that this fat is closer to a carbohydrate and has a declared calorific value of 5 cal g^{-1}.

Salatrim (Benefat™) is produced by Nabisco and contains a mixture of long and very short chain fatty acid triglycerides. This too is said to have a calorific value of 5 cal g^{-1}. Both these fats can only be used in a very limited number of countries.

REFERENCES

1. G. Talbot, 'Chocolate Temper', in S.T. Beckett (ed.), 'Industrial Chocolate Manufacture and Use', 3rd Edition, Blackwell, Oxford, UK, 1999.
2. G. Ziegleder, 'Verbesserte Kristallisation von Kakaobutter unter dem Einfluss eines Scherge Falles', *Int. Z. Lebensm. Techn. Verfahrenst.*, 1985, **36**, 412–418.

Chapter 7

Manufacturing Chocolate Products

Having made the liquid chocolate it is then necessary to turn it into a solid bar, which may or may not have a centre of some other material, such as wafer, biscuit or fondant *etc*. First of all, however, it is essential to ensure that the fat sets in the correct crystal form using a procedure called tempering. A simple chocolate bar can then be made by pouring the tempered chocolate into a mould. Other moulded products have a chocolate shell surrounding a solid or semi-solid centre, or in the case of Easter eggs this centre is left hollow. These products are produced by a process known as shell moulding.

In Roald Dahl's famous story of Charlie and the Chocolate Factory, there are chocolate waterfalls. These only actually exist in miniature, in a machine called an enrober. This produces a curtain of chocolate about 3 cm high which falls over the sweet centres as they pass through it on a wire belt. Many well-known products such as Mars Bar, Lion Bar and Crunchie are made in this way.

A third way of coating centres is called panning. This is used to cover hard centres such as nuts and raisins with a chocolate coating.

Whichever form of processing is used, the chocolate must be allowed to set, so that it can be handled and packed. This must be done correctly, otherwise the chocolate will rapidly become white, with one of two forms of bloom.

TEMPERING

Liquid Chocolate Storage

This is the process which pre-crystallises a small amount of the fat in the chocolate, so that the crystals form nuclei, which help the fat set rapidly in the correct form. The actual amount of fat that it is necessary to crystallise is uncertain, but is probably between 1% and 3%.

The liquid chocolate leaves the conche at a temperature normally above 40 °C. If it is to be processed in the same factory it is then kept in a storage tank until required. These tanks can be 20 tonnes or even more and must be able to be stirred, heated and kept in a lower humidity environment.

If the chocolate is not stirred for a long period some of the fat separates on top, leaving a thicker chocolate in the bottom of the tank. The temperature is maintained at about 45 °C. Keeping it for longer periods at higher temperatures would cause it to change its flavour, whilst the proteins, in milk chocolate, would tend to aggregate making it become thicker. A much lower temperature would start the crystallisation process, and there would be the risk of the tank becoming solid. When describing the transport of cocoa beans it was noted that if the relative humidity was above the equilibrium relative humidity (ERH) of the beans, then they would take up moisture. This is equally true for the liquid chocolate. Here the ERH is between about 35% and 40%, so if the relative humidity around the storage tank is higher than this, the chocolate will take up water. Only a very small amount of water is required to start sticking the sugar particles together and increase the viscosity, making it difficult to process.

Tempering Machines

The tempering machine must first of all cool the chocolate down so that crystals can start to form. Chocolate is a very poor conductor of heat, so for it to cool quickly it must be mixed well, so that all of it comes into contact with the cold metal surface of the tempering machine. These machines are a type of heat exchanger, which heats and cools the chocolate as it passes through it. A typical chocolate temperer is illustrated in Figure 7.1.

The centre column is a rotating shaft to which is attached a series of discs, or scrapers. The walls have bars or discs fitted to them, to ensure that chocolate has to flow alternatively to the wall and to the centre and can't just pass straight through the machine. The faster the shaft turns the higher the shear rate and the faster the crystallisation. In many temperers this is between 3000 and 8000 s^{-1}. The limit is set by the size of the motor and the heat developed by the mixing, which would start to melt the crystals.

The outside surface must be carefully temperature controlled and this is used to determine the degree of temper developed. The temperature is normally controlled within set groups of plates, called zones. Many machines have three or four zones, although many more are possible.

The first zone cools the chocolate to a temperature where crystals can start to form. The second takes it much lower so that Form IV and Form

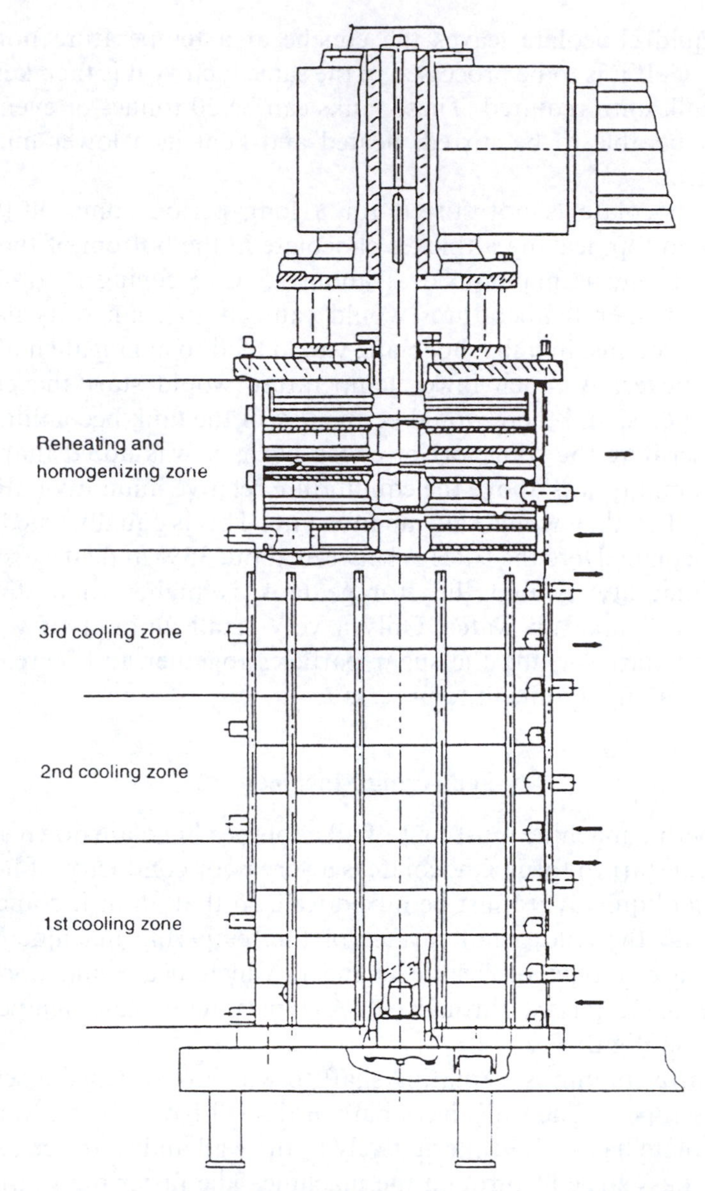

Figure 7.1 *Schematic diagram of Sollich Solltemper MST-V tempering machine* (Nelson[1])

V crystals are also present, and during which the chocolate is highly sheared. The final stage raises the temperature to about 30 °C, which melts out many of the unstable crystals.

The viscosity of chocolate, like most other substances, falls as the temperature rises. The thinner the chocolate the easier it flows into a mould or around a sweet. It is therefore better to process the chocolate at as high a temperature as possible, without melting out the seed crystals. When they are newly formed the crystals are small and are easily melted. If the chocolate is stirred and slowly heated, these crystals become more stable and have a higher melting point. It is for this reason that some tempering machines have an additional stage, where the chocolate is sheared more slowly and the crystals are allowed to 'mature'.

Temper Measurement

When the chocolate leaves the tempering machine it is important for the chocolate manufacturer to be sure that there are enough crystals of the correct type to make the chocolate set properly. In the laboratory it is possible to use X-rays to determine exactly the type of crystals present. This is not only very expensive and slow, but the sugar is often removed first as its X-ray peaks would dominate the spectrum. This may mean that the sample has changed by the time the measurement is taken, which makes it totally impractical for an industrial process.

The differential scanning calorimeter (DSC, see Chapter 8) is able to be used to provide a lot of information about the crystals present in the chocolate, but requires very careful sample preparation. This type of machine can be used for investigative work, but is too expensive and complicated for routine quality control.

In Chapter 6 in the section about solid fat content, it was noted that fats sometimes gave a cooling effect in the mouth because the energy required to melt them – the latent heat – was very much larger than the energy required to change its temperature by 1 °C – the specific heat. When solidifying chocolate this is also true, except that the latent heat is given out when the fat changes from a liquid to a solid, not taken in, as is the case in the mouth. The manufacturer can use this effect to determine whether the chocolate is correct, by measuring a cooling curve.

This can be done very simply and cheaply by the type of device shown in Figure 7.2. Typical cooling curves produced by this instrument are shown in Figure 7.3. The instrument is made up of a metal tube with a cup in the top to hold the chocolate sample. A thermometer probe fits through a cap, which fits tightly on to the cup. The temperature recorded by this probe can be recorded on a chart recorder as shown, or on a computer database. The tube fits through a holder in the thermos flask, which keeps it in position, with the cup above the water level in the flask. A mixture of ice and water is placed in the flask and then the tube put in position. Some of the tempered chocolate is poured into the cup and the thermometer and cap pushed on top. A plot is then made of the temperature against time.

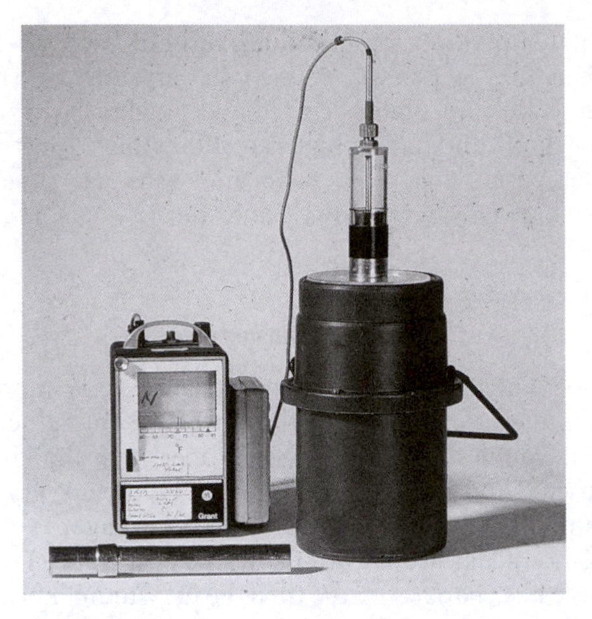

Figure 7.2 *Picture of temper meter*

Initially the temperature drops uniformly but slowly. If the chocolate is tempered correctly, there is enough seed distributed throughout the chocolate to make it set very quickly. As it does so the latent heat is given out, and this off-sets the cooling of the ice so the temperature remains constant for a significant time, as is shown in Figure 7.3a.

If not enough seed crystals are present it takes longer for the chocolate to set. There is, however, more fat to set and so more latent heat to come out. This means that there is a longer initial drop in temperature, but then so much latent heat is released that the temperature of the chocolate rises again. It eventually falls again once the latent heat has come out. This is shown in Figure 7.3b.

If almost no seed crystals are present, or indeed far too many (so that a lot of the latent heat has already been removed) a curve like Figure 7.3c is obtained, which means that the chocolate is not going to set correctly. Here mainly specific heat is being removed. The latent heat comes out slowly and in the case of the uncrystallised chocolate will be less, because the more unstable crystals will be formed with their lower energy state.

The temperature at which the inflection on the curve occurs, point I on the curves in Figures 7.3 a and b is also very important. The higher the temperature the more mature the crystallisation and the higher the temperature at which the chocolate can be used for moulding or enrobing.

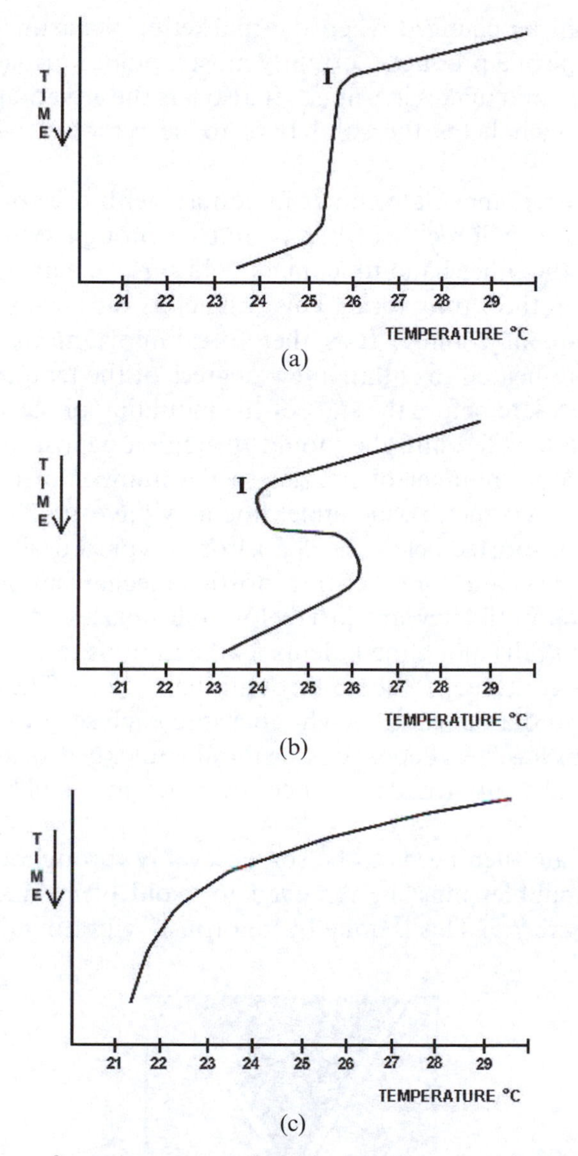

Figure 7.3 *Curves from temper meter for:* (a) *correctly tempered chocolate;* (b) *under-tempered chocolate;* (c) *over-tempered chocolate*

MOULDING

Solid Tablets

This is the simplest method of forming chocolate and is what is used to make the simple tablets. For many years metal moulds were used, but these were heavy, noisy and rather expensive. The latter is very important when a moulding line can contain as many as 1500 moulds, which must all be changed when the marketing department alters the weight or shape of a product. Currently most moulds are made of plastic, which is lighter and makes less noise. It also has the advantage that it can be twisted, which helps the solid bars to be removed when sticking occurs.

If the tempered chocolate comes in contact with a warm surface the crystals within it will melt, so there won't be enough seed for it to set properly. On the other hand touching a cold surface may cause some of the fat to set in the wrong form. This will act as the wrong type of seed during subsequent cooling. It is therefore important that the empty moulds are pre-heated to within a few degrees of the temperature of the tempered chocolate before the start of the moulding process.

The chocolate is fed into the mould through a depositing head. This contains an equal number of nozzles to the number of indents in the moulds below. A 1 metre wide tablet line may have 10–20 nozzles. The moulds are transported below the depositor (a typical design is shown in Figure 7.4). The chains, or other transporting mechanism, sometimes lift up the moulds until they are just below the nozzles. A rod of liquid chocolate then falls along the indents as the moulds move forward and the chocolate continues to come through the nozzles. These depositors are designed so that an exact weight goes into each sector of the mould. As soon as this has been deposited, the mould quickly drops to its former level. This breaks any strands of chocolate which may still be attached to the nozzle.

The chocolate then needs to be spread evenly throughout the mould and any air bubbles must be removed to avoid blemishes of the type shown in Figure 7.5. This is done by vigorously vibrating the mould. In

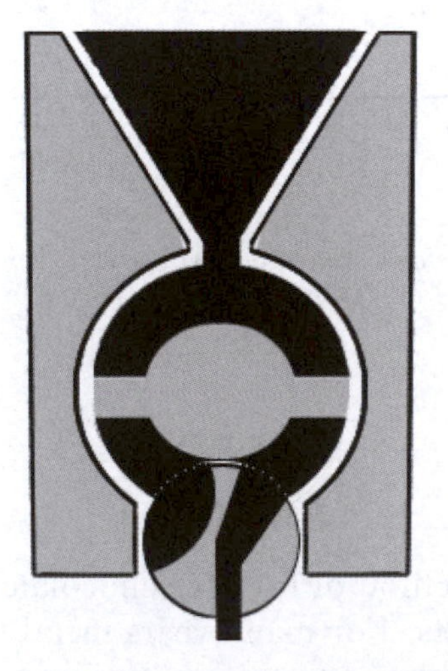

Figure 7.4 *Diagram of piston-type depositor to put a measured amount of chocolate into a mould*

Figure 7.5 *Picture of hole in chocolate caused by an air bubble*

Chapter 6 chocolate was stated to be a non-Newtonian liquid in that it had a yield value. This means that energy must be applied to start it moving. This yield value will also tend to stop bubbles rising through the chocolate. Provided the shaker has the correct frequency and amplitude, however, it can turn a non-Newtonian liquid into a Newtonian one and therefore make the bubbles come out more easily.

When the liquid chocolate is at rest, the particles are practically in contact with one another and so it is difficult for them to move (Figure 7.6a). The vibration provides the energy to separate them, thereby lowering their resistance to movement and the yield value. When a bubble rises it does so very slowly, so we need to know what the viscosity appears to be at a low shear/flow rate. Figure 7.6b shows a plot of the apparent viscosity at a low flow rate against the frequency of the vibration for three different amplitudes.[2] This shows that we must vibrate this particular chocolate at more than 10 cycles per second and with an amplitude of at least 0.2 mm. If we do this the chocolate has one seventh of its viscosity when it is vibrated at 30 cycles per second than when it is unvibrated. Low vibration rates do very little, so the manufacturer is wasting his money in installing the vibrator and also on the energy involved. Very much higher rates do not improve its efficiency. This thinning of the chocolate by vibration only takes place when the vibrator is on and stops as soon as the mould leaves it.

Chocolate Shells

Many moulded confectionery products contain a centre which contrasts with the chocolate in terms of taste or texture, *e.g.* caramel, fondant or praline. Others like Easter eggs are just hollow shells. Both can be made

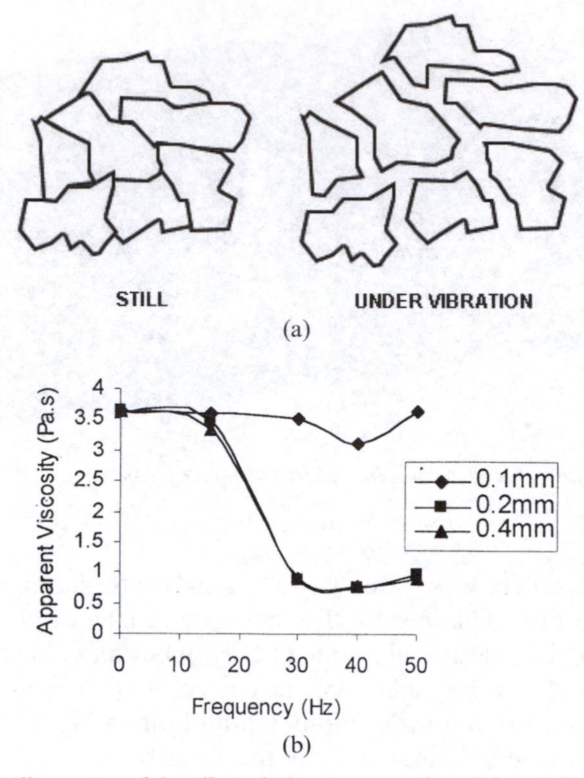

Figure 7.6 (a) *Illustration of the effect of vibration upon the solid particles in a suspension;* (b) *the effect of vibrational frequency, at three different amplitudes, upon the apparent viscosity of a milk chocolate*

on shell moulding lines. A schematic diagram of the procedures involved is given in Figure 7.7.

The first part of the process is the same as for making solid tablets. However, instead of letting the chocolate fully set, the moulds are passed through a short cooling tunnel, so that only the outside sets. The mould is then turned upside down and vibrated. The chocolate initially sets near the mould wall, but should be still liquid in the centre. This liquid part runs out when the mould is inverted leaving the outer shell.

There are several critical features to this procedure. Firstly the chocolate viscosity must be correct, in particular the yield value. If it is incorrect, for instance too high, none of the chocolate may come out. If it is too low, too much may come and the shell will be too thin. Incorrect viscosities also lead to a non-uniform thickness of the shell. Thin sections may allow the centre material to leak from the final product (Figure 4.3), or be very easily broken in the case of Easter eggs. Secondly the cooling must be correct as it has to be long enough to let the shell form, but not so long that most sets or that contraction occurs and the shell falls out too. As with the removal of bubbles from solid tablets, the shaker

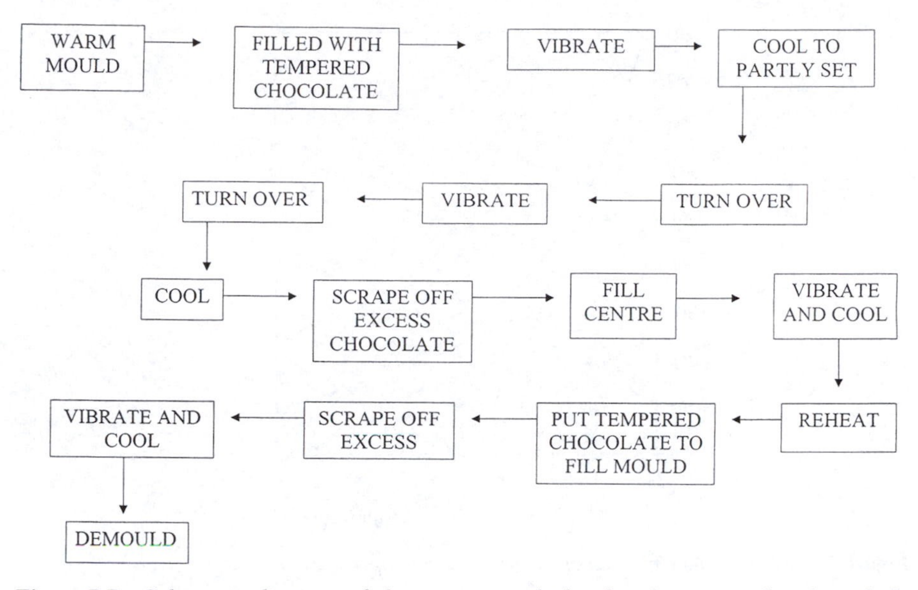

Figure 7.7 *Schematic diagram of the processes which take place on a chocolate shell-moulding production line*

frequency and amplitude must be correct in order to obtain good weight control of the shells.

Once the centre chocolate has run out, the moulds are turned back over, so that the chocolate shell can be filled. In the case of hollow products, just the mould is further cooled, so that the chocolate sets fully and contracts. The mould can then be turned over again and the shell will fall onto a conveyor belt ready for packing. Getting all the half shells for Easter eggs the same weight is not easy, so some packing lines are constructed such that the shells are weighed and divided into three types, heavy, average and light. They are then packed as two average halves or a heavy with a light one.

There are other ways of making hollow eggs and figures. These include book moulds and spinning. In the first technique, two moulds are hinged together and the shell moulding procedure is carried out as normal. The rims are then melted and the hinge brought over like a book to bring the two parts together.

For spinning, the two halves of the mould are clamped together and a measured amount of tempered chocolate is fed into it. The mould is the put on a rotary arm, which turns and rotates so that all the inside is uniformly covered with the setting chocolate. Once it is hard enough, the mould is opened and the hollow figure removed.

When filling the shell, it is important that the centre material does not melt the chocolate. This is relatively easy for fondants or fat based fillings such as pralines, which can be relatively fluid at about 30 °C. With

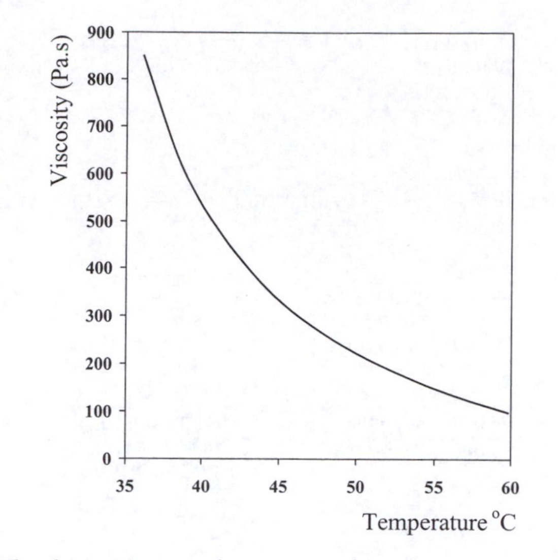

Figure 7.8 *The change in caramel viscosity with temperature*

caramel this is normally much more difficult. Here the viscosity changes very rapidly with temperature (see for example Figure 7.8). It is necessary for the caramel to fill the shell and form a flat top, as any protruding strings or 'tails' may penetrate through the chocolate base (see Figure 7.9). The chocolate normally acts as a moisture barrier to stop the centre drying out (or picking up moisture in the case of centres like wafers), but these tails provide a path for moisture transfer, which will shorten the shelf-life of the product. In addition any stickiness on the outside may cause problems during wrapping. This means that the caramel must be added as hot as possible, to make it flow as easily as possible, and yet not so hot that it will melt the chocolate shell.

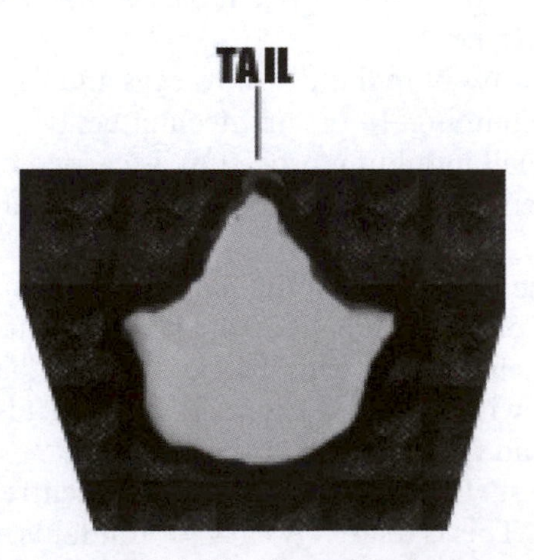

Figure 7.9 *Illustration of a caramel 'tail' within a chocolate coated sweet*

Figure 7.10 *Picture of scraper blade used to form a uniform back on the sweets*

Once the centre is in place then a chocolate base must be put on the shell. This is often done by having a scraper blade pressing against the mould as shown in Figure 7.10. Tempered chocolate is poured onto the moulds in front of the blade and forms a rotating roll in front of it. This fills in any remaining indents or holes, whilst the blade removes any excess chocolate.

The moulds are then cooled in a cooling tunnel, to fully set the shell and base, before they are turned over again and the sweets dropped on to a belt, which carries them into the packaging room. Because the outside fat has set in contact with the mould surface, this makes it smooth and very glossy. In any chocolate selection box, the moulded sweets are normally much shinier than those made using the enrobing process.

ENROBERS

In this case the centre material, be it nougat, biscuit, fondant or caramel, is first of all made separately and then placed on a belt, which will take it through the enrober. The aim is to get all the surfaces, including the base and ends, coated uniformly with chocolate. Typical examples are Mars Bar, Lion Bar and Cadbury's Crunchie.

As with moulding it is necessary to use tempered chocolate. The tempering unit may be located in the enrober line or placed next to it with the chocolate being transferred in jacket-heated pipes.

A typical enrobing system is illustrated in Figure 7.11. The centres are placed on a continuous moving wire chain belt (A) as is shown in Figure 7.12, which transports them underneath the chocolate 'waterfall' (B).

Figure 7.11 *Picture of sweet centres going into a chocolate 'waterfall' in an enrober*

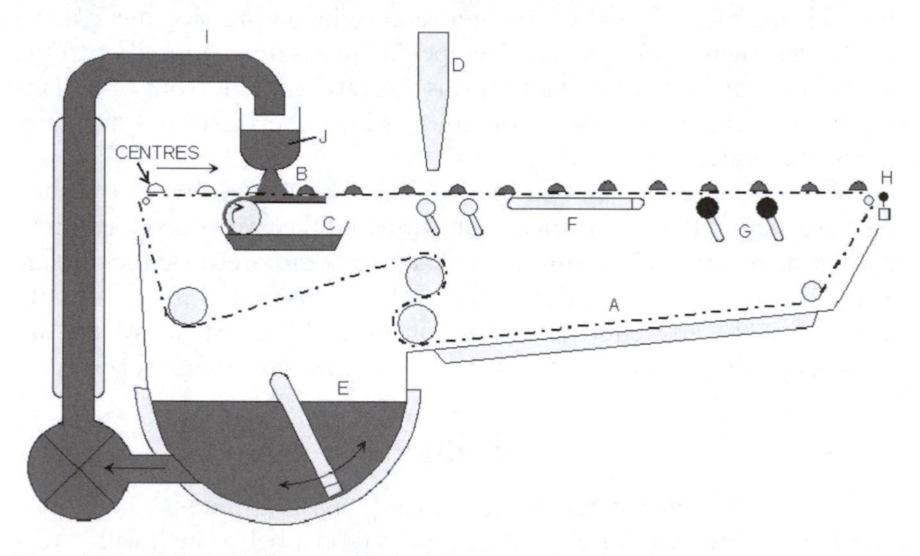

Figure 7.12 *Schematic diagram of an enrober*

Just beneath the belt at that position is the bottoming or surge trough (C). This retains the chocolate that falls through the chain belt and recirculates it, by means of the end roller over the plate, which is just below it. This chocolate moves along with the centres and pushes them up slightly, forming a layer of chocolate on the under side. Sometimes there are two 'waterfalls' within a single enrober, but at other times two enrober systems are used one after the other. This is particularly useful when the product has an uneven surface, *e.g.* crispies sticking out. The first coating will have a low yield value so that it flows into all the crevices and provides a good moisture barrier for the product. The second may have a high yield value, so that the chocolate 'stands up' giving the product a more rugged appearance.

Having poured chocolate over the centres it is then necessary to make sure that each item has the correct weight and appearance. This is done by a series of devices, with intensity control that can be varied according to the product being made.

The first is a jet of warm air (D) which blows the excess chocolate from the top and sides of the sweet so that it passes through the belt back into the base trough of the enrober (E). This tends to leave a ruffled wavy surface, especially if the chocolate has a high yield value. This is then smoothed and additional chocolate removed by the shakers (F). As was described earlier, these should be operated at the correct frequency and amplitude to overcome the yield value.

The next part deals with the bases. The chocolate was put on in the bottoming trough, but now may be too thick, or indeed be uneven and have uncovered areas. This is dealt with by the grid-licking rolls (G). Depending upon the design of machine these normally vary in number from one to three and may operate at different speeds and distances from the wire belt. This enables chocolate to be put on or removed from the base. Any remaining chocolate on the rolls is scraped off by a knife and once again falls back into the base trough.

The next section of the belt can be used to decorate the top of the sweets. For many years this was done by hand using marking forks of the type shown in Figure 7.13. Now most products are marked by rotating

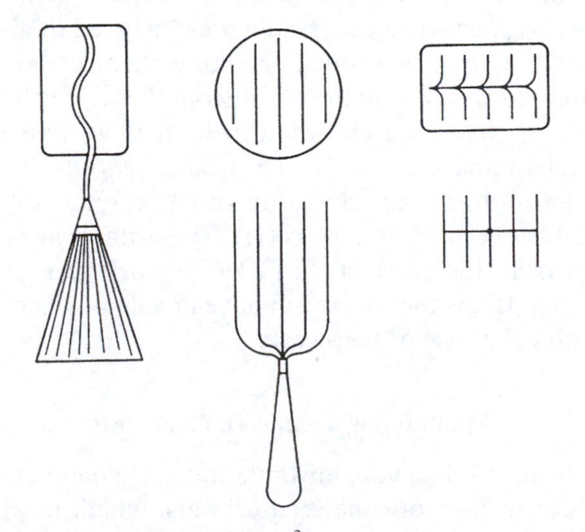

Figure 7.13 *Hand-decorating tools* (Nelson[3])

Figure 7.14 *Chocolate with 'tails' on a cooling tunnel belt*

roller, which just touches the top, or by programmable moving nozzles, which are able to put a wide variety of markings on top, in the same, or even a different coloured chocolate.

Finally the product passes onto another plastic belt which passes through the cooling tunnel. This belt is often embossed with the name of the manufacturer or the product brand. As the chocolate is still semi-liquid the base then takes up this lettering. This is another way, other than the gloss of determining whether a product has been moulded or enrobed, as an enrobed sweet will have any detailed lettering on the base, with less distinct top markings, whereas moulded sweets are able to have much greater detail on the top and sides, but the base is normally plain.

Frequently this plastic belt moves at a greater speed than the wire one. This separates the rows of product, making them easier to select for packing. It does, however, pull the sweet from the wire belt often leaving long thin tails of the liquid chocolate behind them (see Figure 7.14). When these tails remain attached, the product is difficult to pack and often has to be sold as a mis-shape. In order to remove them, a small roller is placed between the two belts (H). This rotates at high speed and pulls the tails from the product. In order to work properly it must be correctly located. If it is too low it will have no effect, whereas too high a setting will mark the base of the sweets.

Maintaining Tempered Chocolate

Tempered chocolate is in a very unstable state. The temperature must be low enough not to melt out the seed crystals, which means that other crystals will be forming. Eventually the chocolate will become thicker and after a very long time will solidify. In the enrober the tempered chocolate is fed into the base trough (E in Figure 7.12). From here it is pumped up a pipe (I) into the flow pan (J). This pan is across the full width of the wire belt and has one or two slits along the bottom, which form the 'waterfall'. As was described earlier much of this chocolate passes through the wire belt due to blowing or shaking or simply because it was poured over an area of the belt where no sweet was present. Eventually this chocolate would become thick and overtempered.

Newly tempered chocolate is added to the base trough to replace that which is removed by the product, but normally this is not enough to stop this thickening occurring. In order to maintain stable conditions, therefore, some of the chocolate is continuously removed from the enrober base tank. It is then heated to between 40 °C and 50 °C to remove all the fat crystals, and then returned to the input of the tempering machine.

SOLIDIFYING THE CHOCOLATE

When the chocolates leave the moulding plant most of the fat is still in the liquid state. In order for it to be firm enough to handle and package, most of this must solidify in the correct crystalline form. This requires the removal of a large amount of latent heat and a relatively smaller amount of specific heat. The chocolates are probably already at a temperature several degrees lower than 30 °C and when they are ready for packing they will be at around room temperature, a drop of less than 10 °C. The specific heat of chocolate is about 1.6 J g^{-1} $°C^{-1}$, so about 16 J must be removed from each gram of chocolate. The latent heat, on the other hand is 45 J g^{-1}, so 45 J is removed for each gram, giving a total of just over 60 J to solidify and cool that amount of chocolate.

There are three ways that a body can loose heat, *i.e.* conduction, radiation and convection. In conduction the heat flows through the material directly in contact with that being cooled. In this case the chocolate is only in contact with the plastic mould or belt. Both these are very poor conductors of heat, so very little heat energy can escape by this method, although a cold surface below the belt will help crystallise thin flat products. Radiation heat transfer takes place at a rate determined by the fourth power of the temperature difference between the object being cooled and its surroundings (ΔT^4). In this case the sweet is at about 25 °C. If we assume that the surroundings are at 0 °C, which, as will be shown later, is lower than may be desirable, then the heat transferred to this low temperature absorber will be 126 W m^{-2}.

The third method is to blow cold air over the product. Some of the heat is given up to this air, which then moves away and is recooled. If the air is cooled to 0 °C and blown over the product at 240 m min^{-1} then heat can be transferred at 630 W m^{-2}, that is more than 5 times faster than by radiation (Nelson[3]).

Low temperatures, however, may give rise to two problems. Firstly they may cause the fat to set in the wrong crystalline form. This will cause the product to bloom very quickly and moulded chocolates not to contract properly and so become difficult to demould. Secondly moisture in the air may condense on cold surfaces and then drip onto the chocolate. This will dissolved some of the sugar from within it. When the chocolate is rewarmed ready for packing, the water evaporates again, leaving a white powdery surface. This looks very like fat bloom, but is in fact sugar and is therefore called sugar bloom. If possible the temperature of the surrounding tunnel should be maintained above the dew point i.e. that temperature at which moisture just starts to condense.

Coolers

Many coolers are just long tunnels, with blowers and coolers at intervals, so that it can be divided into different temperature zones. A diagram of the cross section at such a position is given in Figure 7.15.

The initial cooling, particularly for enrobed products, is fairly gentle. This is then followed by the coldest part, which corresponds to the position where most of the latent heat is given out. This is normally about 13 °C, but lower temperatures are possible, provided the air is moving rapidly to prevent any condensation. The temperature is then raised slightly before the product enters the packing room. This is because, if the product surface temperature is lower than the dew point temperature in this room, moisture will condense on the chocolate causing blemishes, or maybe sugar bloom.

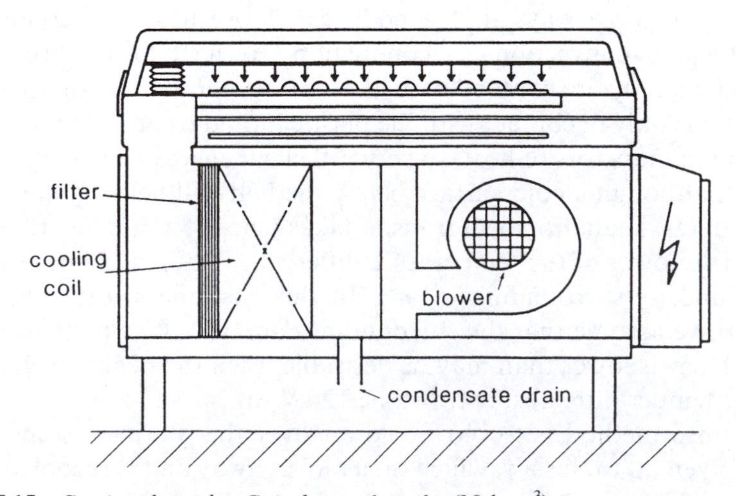

Figure 7.15 *Section through a Gainsborough cooler* (Nelson[3])

Where space is limited, and especially for moulded products, multi-tier coolers are used. Here the product goes backwards and forwards through different temperature zones. A three tier cooler, therefore, has a cooling time six times that of an equivalent length tunnel cooler.

The main problem with the multi-tier system is that the product must be kept horizontal as it is lifted from one level to the next one. This is sometimes achieved by a continuous chain driven system, which holds the moulds or plastic trays, on which enrobed sweets are placed.

The time it takes for a product to set depends not only on the amount of crystals already in the chocolate, but also its type and amount. A large block of chocolate is naturally going to take much longer than some small morsels being made for cookies. Normally, however, it takes between about 10 and 20 minutes to obtain a good quality product.

PANNING

This process is used to make the small rounded chocolate items often sold in bags or tubes. These can be divided into two types, the chocolate coated varieties, normally containing a nut or dried fruit and the chocolate centred sugar coated products such as Smarties and M&Ms. Both can be made using open rotating bowls or pans (Figure 7.16), often constructed of copper and which have temperature and humidity controlled air blowing into them through a pipe. These have often been replaced on the industrial scale by large rotating drums (Figure 7.17),

Figure 7.16 *Picture of open pan for chocolate or sugar coating*

Figure 7.17 *Picture of rotating drums used for chocolate or sugar coating*

which resemble very large washing machines and can hold more than 2 tonnes of product at a time. These too have temperature and humidity controlled air and the facility to spray liquid ingredients on to the tumbling centres within the drum.

Chocolate Coating[4]

For this type of product it is important to spend some time selecting good centres to coat with chocolate. It is better if all of them are of a similar size. As all the centres tumble over each other in the rotating pan or drum (Figure 7.18), further segregation takes place, with the big ones going to certain areas and the small ones to areas where they are much less likely to get coated with chocolate. This means that at the end of the process there is an even bigger range of sizes (see Project 2, Chapter 10).

The shape is also important. Where possible sharp edges should be avoided as these are difficult to coat and will often show through the chocolate. Slightly convex shapes are also better than concave ones. As is shown in Figure 7.19, convex ones can only touch at a single point and so are easily separated by the tumbling action. The liquid chocolate, on the other hand, can get within a concave impression and form a sticky layer, which will attach it to the next particle. This will result in a lot of doubles or multiple pieces within the product.

The temperature of the centre is also important. Colder centres will help the chocolate to set more rapidly, but too cold a temperature will

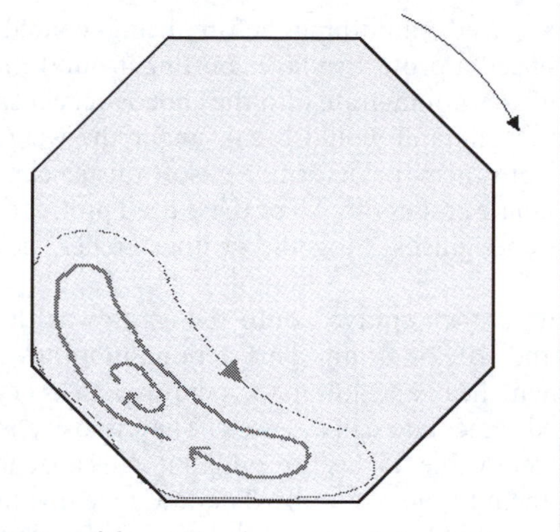

Figure 7.18 *Illustration of the rotational movement during panning*

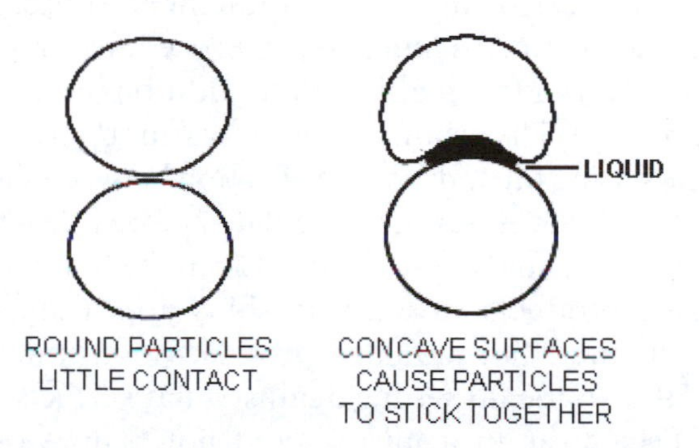

Figure 7.19 *Illustration of the effect of surface curvature on the formation of agglomerates during panning*

cause it to be uneven and to crack off at a later stage. In addition the size of the centre will change with temperature according to its coefficient of cubical (or volume) expansion (γ). This can be expressed by the equation:

$$V_t = V_o(1 + \gamma t) \tag{7.1}$$

where V_t = volume at temperature t, V_o = volume at 0 °C.

The coefficient is therefore the increase in volume when 1 cm^3 of a substance is raised in temperature by 1 °C. This is different for different substances and may be much higher for some centres than for the surrounding chocolate. This means that the centres should be coated, stored and sold at a relatively constant temperature, otherwise the centre will expand and crack the outer shell, which is unable to expand by the same amount. This is illustrated in Project 17 in Chapter 10.

Many centres need smoothing before being coated in chocolate, whereas others need a protective layer putting around them, so that the oil they contain does not migrate into the chocolate, causing it to bloom. Yet others are fragile and would break under the weight of the other centres falling onto them, or deform, *e.g.* soft raisins change their shape so that the chocolate cracks off. All of these need pretreating often with a sugar (sucrose and glucose) together with starches, gelatines or gum Arabic.

The chocolate is then sprayed onto the centres as they tumble over each other in the pan or drum. This action smoothes it out over the surface. The high shear and lower temperatures also crystallise the fat and untempered chocolate can be used. The viscosity is important, in particular the yield value. If this is too high it will cause uneven coatings and even stick them to the pan walls. Too low a yield value will not stick to the surfaces and will leave areas of the centre uncoated.

The first few layers of chocolate must be put on very carefully. As soon as a uniform layer is seen on the centres, the temperature can be reduced to help it set. It is then warmed again and this procedure is repeated often for three to five times. After this the chocolate additions and cooling can take place at the same time. The product actually warms up from the latent heat given out by the setting fat and from the friction between the falling pieces. Too cold air will, however, set the chocolate before it has been smoothed by the tumbling action and cause the product to be bumpy.

Many of these products have a very shiny appearance. It was noted earlier in this chapter that moulded products look much glossier than enrobed ones, due to the fat setting against a flat surface. Gloss is due to light being reflected, as in a mirror, and not being absorbed by any surface scratches or blemishes. Some moulded products in fact look relatively dull owing to the scuffing they receive during the wrapping procedures. This would happen for a lot of these products too, if it wasn't for the protective coating that is used to polish or seal them. A very common coating is made from shellac, the refined resinous secretions from the lac insect. This is diluted with alcohol and can be sprayed onto the product in the pan, in the same way as the chocolate. This solution will, however, interact with the chocolate giving a poor product unless an inner coating of sugar syrup, glucose and colloid (gum Arabic, starch *etc.*) is used.

Sugar Panning

This process is used to coat the chocolate or similar centres with sugar. For chocolate panning, temperature control is used to harden the coating; for sugar coating it is moisture reduction. The coatings are

Figure 7.20 A method of forming chocolate centres for sugar panning

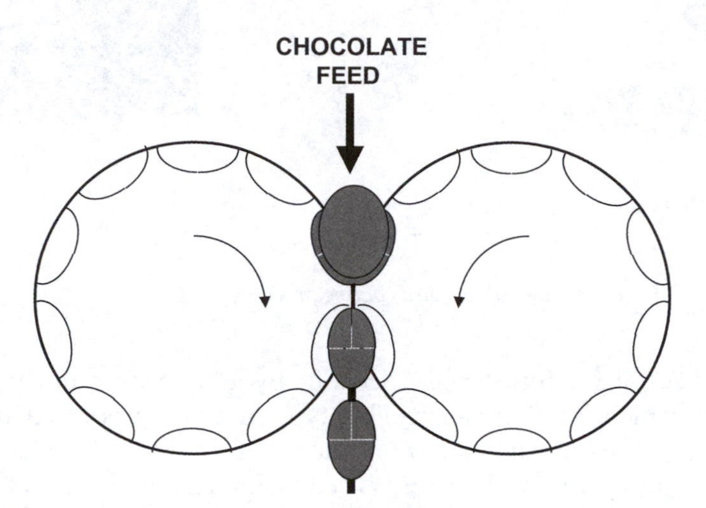

Figure 7.20 *A method of forming chocolate centres for sugar panning*

added in water based solutions and dry air is blown over them so that it evaporates, leaving very fine crystals.

The centres can be made by feeding tempered chocolate in the gap between two cooled rollers (see Figure 7.20). These rolls have matching indents, so that a thin web comes out of the bottom holding together all the centres. This web is removed and the centres fully solidified ready for panning.

The pan or drum is the same as for chocolate coating. The solution is then sprayed in. There are two types of coating, *i.e.* soft coating and hard coating. In the former (Figure 7.21) a liquid is applied and allowed to

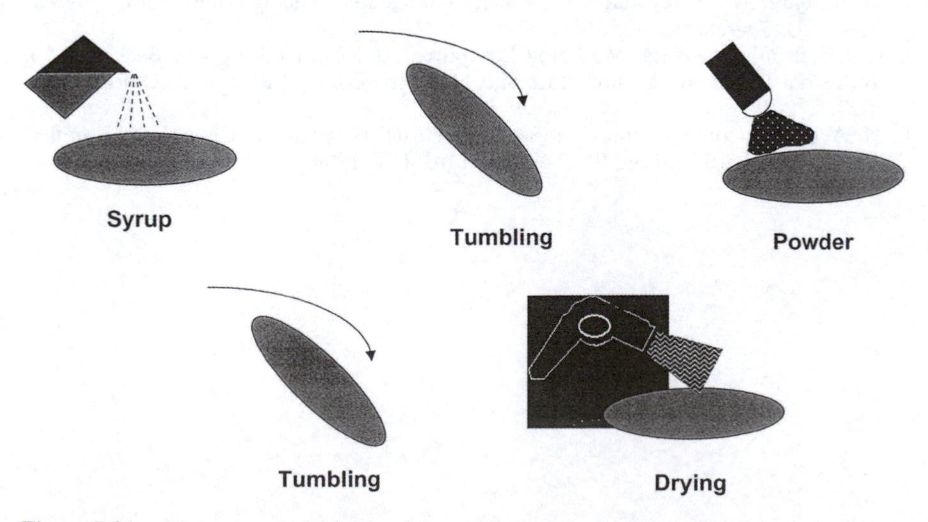

Figure 7.21 *Representation of the soft coating process*

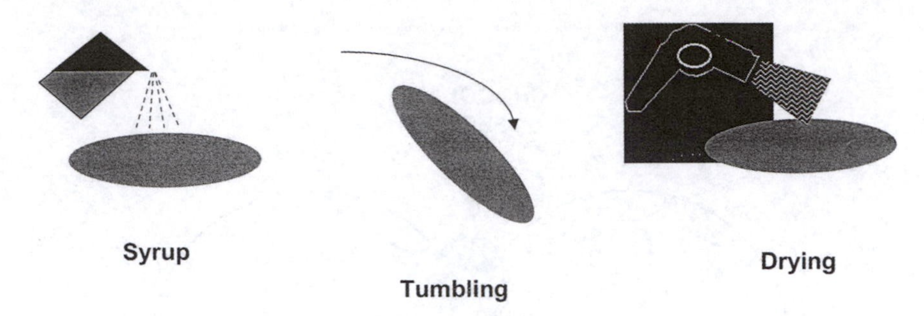

Figure 7.22 *Representation of the hard coating process*

coat the surface before being dried out by adding an absorbing solid material, which is often sugar. This type of coating is often used for jellybean and sugar egg types of product.

Hard coating (Figure 7.22), as is used in Smarties and M&Ms, has the sugar dissolved in the water. In order to minimise the drying time, the concentration is such that it is close to being a saturated solution at the temperature of use. The rate of degree of drying is important in order to avoid cracking and also not to trap in moisture, which will later migrate to the surface and cause blemishes.

Coloured pigments can be added with the sugar and in order to obtain a shiny surface, a wax outer coating may be put on the outside to give a smoother surface.

REFERENCES

1. R.B. Nelson, 'Tempering', in S.T. Beckett (ed.), 'Industrial Chocolate Manufacture and Use', 3rd Edition, Blackwell, Oxford, UK, 1999.
2. M. Barigou, M. Morey and S.T. Beckett, 'Chocolate – The Shaking Truth', *International Food Ingredients*, 1998, **4**, 16–18.
3. R.B. Nelson, 'Enrobers, Moulding Equipment and Coolers', in S.T. Beckett (ed.), 'Industrial Chocolate Manufacture and Use', 3rd Edition, Blackwell, Oxford, UK, 1999.
4. M. Aebi, 'Chocolate Panning', in S.T. Beckett (ed.), 'Industrial Chocolate Manufacture and Use', 3rd Edition, Blackwell, Oxford, UK, 1999.

Chapter 8

Analytical Techniques

A wide variety of analytical techniques is applied within the chocolate making industry. This chapter reviews some of the more widely used ones that are needed to monitor factors such as particle size, moisture, fat content, viscosity, flavour, texture and crystallisation. This is intended to give an insight into the science behind the techniques themselves, rather than to describe a particular instrument.

PARTICLE SIZE MEASUREMENT

Both the size of the biggest particles and the total (specific) surface area are important to chocolate manufacturers. Instruments like micrometers and optical microscopes can measure the diameter of the biggest particles, whereas sieving provides the weight of solid material above a certain size. None of these, however, can record the surface area of the smaller particles. Optical microscopes can measure particles as small as 5 microns, but below this level diffraction of the light makes readings very uncertain. (Diffraction is the bending of a wave motion around the edge of an obstacle – see Figure 8.2). Two types of instrument can be used to measure smaller ones; these are based on electrical conductivity and laser light scattering respectively.

Both methods rely on dispersing the non-fat solid material in a liquid, before the measurement is carried out, so as to separate the individual particles. In chocolate these particles are so tightly packed that these systems would otherwise be unable to differentiate between the different particles. The dispersing liquid must, of course, have no effect on the particles themselves. This means that they must not contain any water, which would otherwise dissolve the sugar. Very often oil or an organic solvent such as trichloroethane is used. A few grams of chocolate are placed in the dispersing liquid. In order to separate the particles some form of vigorous mixing is required, but this too must not be so hard that it breaks the particles. Placing the mixture in a low power ultrasonic bath normally disperses the particles satisfactorily.

The electrical system operates by passing this dispersion through a narrow tube with electrodes on the side. The electric conductivity will be different when there is a particle present from that if there is only the liquid in the tube. In addition, this conductivity will depend upon the particle's volume and electrical conductivity. If it is assumed that sugar, cocoa and milk solids all have a similar electrical conductivity, then by passing particles individually through the electrodes it is possible to obtain a volume distribution. The mathematical treatment assumes that all particles are spherical, in order to obtain an equivalent diameter. This is not exact, but the measurements obtained are sufficiently accurate to be very useful to the chocolate maker. Because the particle size distribution varies from sub-micron to about 70 microns, however, and because the electrode sensitivity depends upon the diameter of the tube, it is often necessary to carry out the measurements with at least two sizes of tubes. This makes the procedure relatively slow, and to a large extent this type of instrument has been superseded by those based on laser light scattering.

The sample preparation is the same as described above. The dispersion is this time circulated through a sample cell, which has a laser beam directed through it (see Figure 8.1). The laser beam is expanded to a diameter of between 5 and 20 mm and then shone through the suspension. In this instrument there is no attempt to measure individual particles, but only the dispersion pattern from all the dispersed particles within the beam. This pattern remains stable even though all the particles within the beam are in motion. The individual particles do, however, diffract the light. Diffraction is due to the bending of a wave motion around the edge of an obstacle. The smaller the aperture or particle

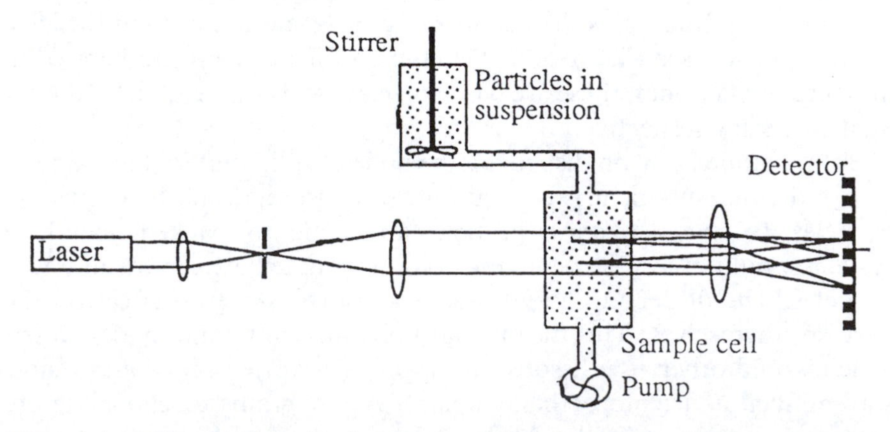

Figure 8.1 *Schematic diagram of a particle size distribution measuring instrument based on laser light scattering*

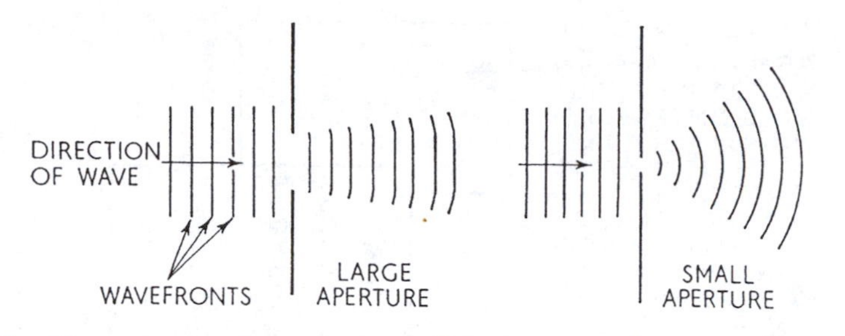

Figure 8.2 *Illustration to show why more light is scattered at a small aperture (or particle) than at a large one (both relative to the wavelength of the light)*

compared with the wavelength of the light (for a helium neon laser it is 0.63 microns), the more it is diffracted (Figure 8.2). A combination of all the different scattering patterns is then imaged by a lens and monitored by a photodiode array. By changing the focal length of this lens, different particle size ranges can be monitored. Typical instruments measure 0.5–90 microns with one lens, but if partly manufactured chocolate is being monitored, *e.g.* from a two roll refiner, the focal length can be increased to change this to say 1–175 microns.

The diffracted patterns are compared with those that would be expected assuming a combination of Fraunhofer diffraction for the bigger particles (⩾40 micron) and Mie scattering for those of a size similar to the wavelength of the light. Particles must, of course diffract/scatter the light individually, and not behave as agglomerates. When the dispersion becomes too concentrated, some of the smaller particles will coincide to give the signal of a bigger one. This means that before taking the measurement the dispersion concentration must be adjusted to give the correct signal.

Once again it is assumed that the particles are spherical to calculate the volume of particles within certain diameter ranges, particle volume distribution curves and even specific surface areas. A typical printout from such an instrument is given in Figure 8.3. Although restricted to particles within the detection range, the normalised data is produced within a few minutes and this technique is therefore very useful for quality control.

MOISTURE DETERMINATION

Chocolate contains approximately 30% of fat, but only about 1% of moisture. Although moisture can migrate through chocolate it is a very slow process, so removing the moisture from large pieces would take a very long time. In order to measure the moisture therefore, the chocolate can either be dispersed in another medium before drying or the water

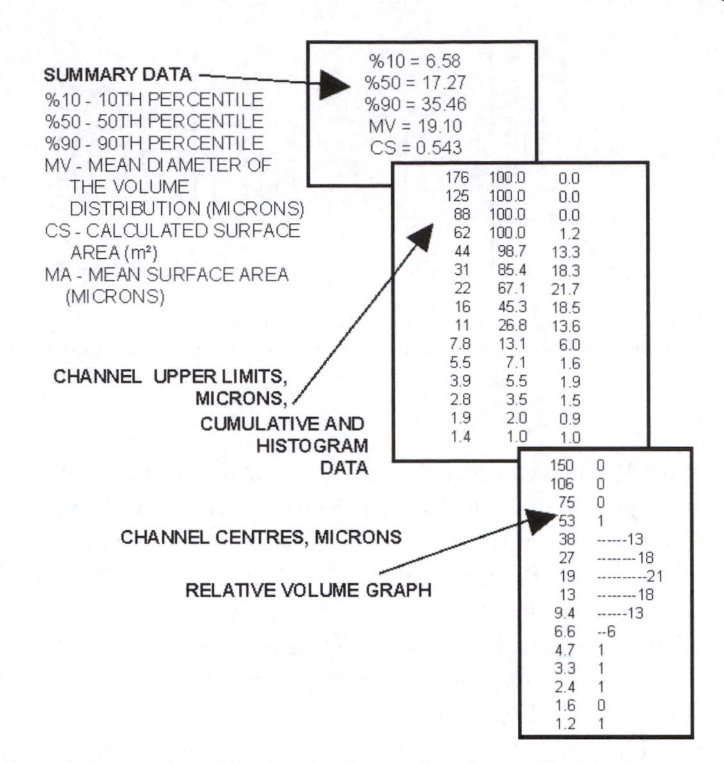

Figure 8.3 *Printout of particle size distribution data for a milk chocolate as measured by a light scattering device*

determined by a chemical reaction, which determines the presence of the H_2O molecule.

Traditionally moisture is determined by mixing a known amount of chocolate with pre-dried sand in a small metal dish, which is then weighed. The dishes are then heated in an oven for a set time and temperature. This may be 98 °C for 12 hours or a lower temperature, for a shorter time but under vacuum, before removing the dishes and reweighing them. Many laboratories have their own standard procedures, which are normally reproducible within that laboratory. Although the loss in weight does not give an absolute value for moisture, as not all the moisture is necessarily removed and a significant amount of other volatile material may have been, it does give a good indication of the 'free' moisture within the chocolate. It is this 'free' moisture that has the very big effect upon the chocolate's flow properties.

An alternative method uses the Karl Fischer reaction to determine the 'free' and 'bound' moisture within the chocolate. That is the moisture that is removed by oven drying together with any that is part of the water of crystallisation, *e.g.* in the lactose monohydrate from the milk. The latter does not play a significant role in effecting the chocolate flow properties and so does not really need to be measured. This technique therefore normally produces a higher measurement for milk chocolate than oven drying, but it does have the advantage of normally being more reproducible when automatic titration systems are used.

The Karl Fischer reaction depends upon the reaction of sulfur dioxide, iodine and water:

$$I_2 + SO_2 + H_2O = SO_3 + 2HI \quad (8.1)$$

This reaction takes place very easily in a solvent composed of methanol and pyridine, according to the equation:

$$I_2 + SO_2 + H_2O + CH_3OH + 3py = 2pyH^+ I^- + pyHSO_3OCH_3 \quad (8.2)$$

This can take place in solvents other than methanol, but the pyridine must always be present. The overall reaction means that one molecule of water reacts with one molecule of iodine.

In the automatic system a pre-weighed amount of chocolate is dispersed in a mixture of formamide, chloroform and methanol in the reaction vessel. This vessel is then sealed to prevent further moisture coming in from the surrounding air. The mixture is then continuously monitored by two platinum electrodes that are placed in it. If any free iodine is present, this will depolarise the cathode and so the current will stop flowing. The potential difference across the electrodes is therefore used to control the titration. Initially Karl Fischer reagent (which includes pyridine) is added slowly via a peristaltic pump, which is able to record accurately the amount that passes through it. When the reaction is complete, and no moisture remains, the current will stop flowing. The instrument can then calculate the percentage of water present, based on the sample weight and the amount of reagent used.

FAT CONTENT MEASUREMENT

Devices based on the reflection of near infrared radiation off the chocolate surface are able to give immediate readings concerning its fat and moisture content. This radiation is, however, affected by other factors such as particle size and chocolate composition, which means that instruments need to be recalibrated for each product. The calibration for fat can be carried out using the traditional soxhlet method to determine its actual content.

The soxhlet method involves using a solvent to dissolve the fat out of the chocolate. The solvent is then evaporated, leaving the fat, which can then be weighed. This analysis can be carried out in a glass system, such as is illustrated in Figure 8.4.

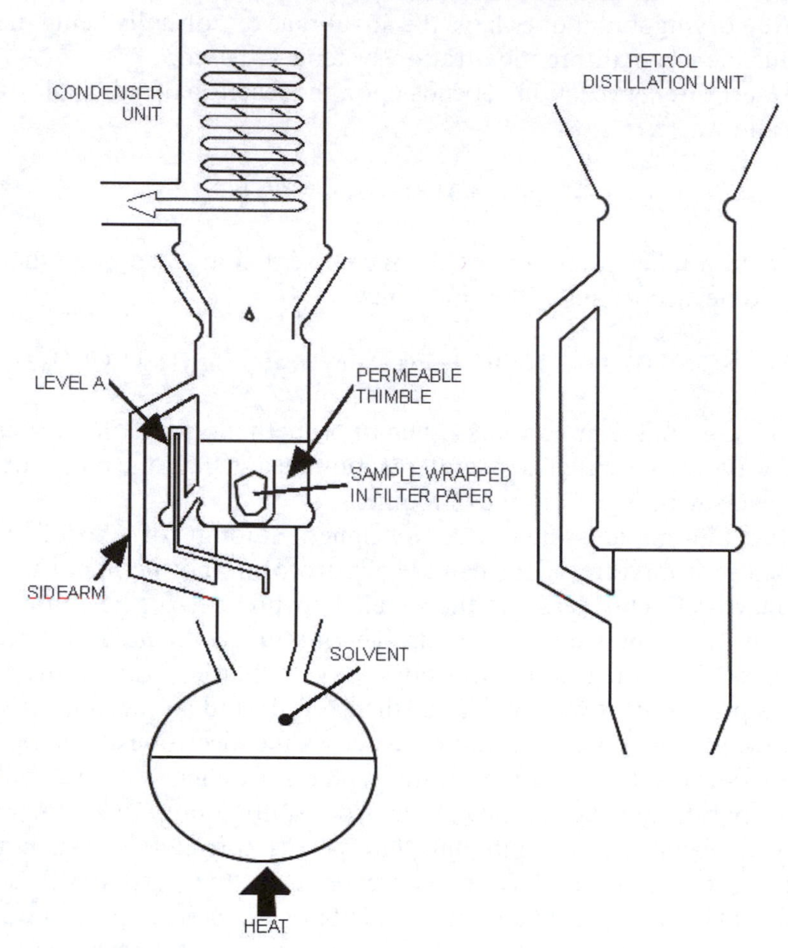

Figure 8.4 *Diagram of soxhlet fat extraction apparatus*

A weighed amount of chocolate is finely divided into small pieces before being wrapped in a filter paper, which is then put inside a permeable thimble container. This is then placed in the central part of the extraction system, the top part of which is cooled by a water jacket. The solvent, normally petroleum ether is poured into the bottom flask where it is heated by an electric mantle. As it boils, the ether vapour rises through the sidearm and condenses in the top section. Here it runs down into the middle container where it comes in contact with the chocolate and dissolves the fat. The fat-containing solution passes through the filter paper and thimble and collects in the container until it reaches level A, corresponding to the top of the siphon system. It is then able to run back into the bottom flask. The solid material within the chocolate is retained within the filter paper and the fat collects inside the flask at the bottom – the mantle is hot enough to boil the ether, but not the fat.

Once all the fat in the chocolate has been removed, normally after about 12 hours, the top sections are separated from the flask. A new section, the petrol distillation unit, is placed on top, and will collect all the petroleum ether and not let it drip back. The fat is therefore left in the bottom of the flask and can be weighed. The whole procedure should take place within a fume-cupboard, with great care being taken to ensure that electrical connections and other objects present, are unable to produce sparks and cause an explosion.

VISCOSITY DETERMINATION

Simple Factory Techniques

In Chapter 5 it was shown that liquid chocolate is a non-Newtonian fluid and so its viscosity can not be described by a single figure. In a simplified form it can be described as having a plastic viscosity and a yield value. Simple instruments do exist which will give single point measurements, two of which are the ball fall and the flow cup, shown in Figure 8.5 and used in Project 5 in Chapter 10.

In the ball fall viscometer, there is a round weight attached to a wire or rod. On the latter are two marks. The ball is dropped into a container of liquid chocolate, normally controlled to be at 40 °C. Accurate temperature control is very important as a small difference in temperature has a big effect upon chocolate flow properties. The time for the distance between the two marks to be covered in chocolate is noted and relates to the viscosity. Normally this measurement is carried out several times, but

Figure 8.5 *Picture of ball fall and flow cup viscometers*

the first reading is neglected. If a bigger ball is used it will fall faster, so a series of measurements can be obtained by using several different sizes, and a flow curve is produced which is related to that chocolate. The rate of fall is, however, relatively slow for most chocolates, and the readings tend to relate to the yield value rather than the plastic viscosity.

In the flow cup, on the other hand, the chocolate is moving much faster, so its results relate more closely to the plastic viscosity. Here the chocolate and cup are warmed to the measurement temperature (usually 40 °C). The cup is held above a balance on which is placed the receiving container. The plug is raised and the time recorded for a set amount of chocolate to run into this container. The thinner the chocolate, the faster it will flow and the shorter this time will be.

The Standard Method

In order to measure the viscosity over a wide range of shear rates (degrees of flow or mixing) a concentric cylinder viscometer is used. This consists of a cup which contains the chocolate and a central bob. This bob normally has either a pointed bottom (DIN) or has a hollow underneath, which is full of air when the viscometer is used (see Figure 8.6). The reason for this is that when the bob or cup rotates the chocolate will try to prevent it doing so. There is in fact the situation shown in Figure 5.2, with two parallel areas a fixed distance apart, moving at a different speeds. The viscosity is directly related to the force the liquid develops to stop this movement. At the base the velocity is not uniform, being zero in the centre and maximum at the edge. This makes calculating the viscosity very difficult. These two designs have therefore been developed so that almost all of the related force comes from the parallel area between the bob sides and the cup and almost none from the base.

It is also very important that the chocolate is uniformly treated within the gap. If the gap is wide and the bob turns, only the chocolate near the

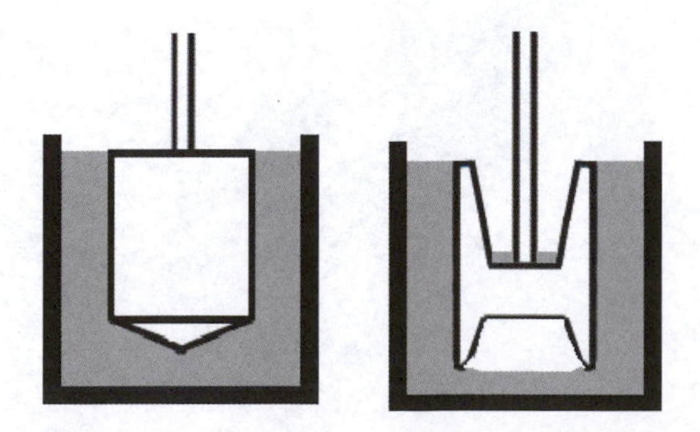

Figure 8.6 *Two different types of viscometer bob that are used to measure the viscosity of chocolate*

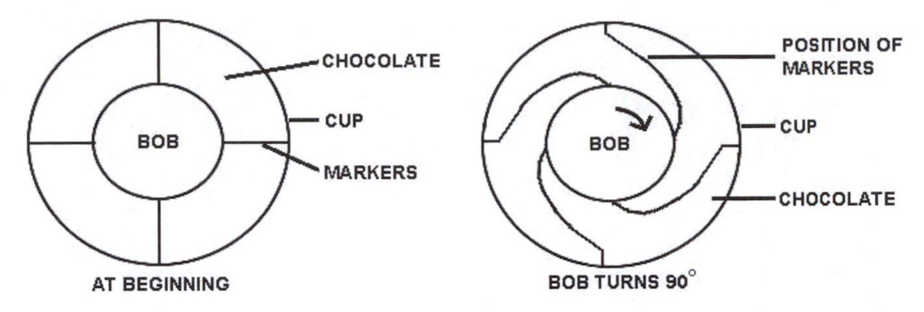

Figure 8.7 *The flow between the cup and the bob of a concentric cylinder viscometer*

bob is going to be effected. The chocolate near the cup wall may even remain stationary (see Figure 8.7). To make matters worse the viscosity of the chocolate varies depending upon how fast it is moving. This means that even in a relatively narrow gap between the bob and cup there may be a range of viscosities across it. In order to get a uniform flow across the gap it must be narrow and usually the ratio of the bob to the cup diameter must be $\geqslant 0.85$.

The chocolate must be free from fat crystals before the measurement is taken. To ensure this, the chocolate is heated to 50 °C and then cooled back down to just above 40 °C before pouring it into the pre-warmed cup. The bob is then put in and turned slowly so that the chocolate comes to a uniform temperature. The speed of rotation is then increased, with readings of the retarding force being taken at intervals. The speed of rotation is also noted, and can be converted into a shear rate (shear rate = relative velocity/distance between the bob and cup). The viscometer is kept at the maximum stirring rate for a short period and then readings are taken at the same points as the speed is reduced.

In some viscometers the cup rather than the bob rotates, and the force on the bob is measured. On others a specific force is applied to the bob and its speed of rotation is measured. For all of these the apparent viscosity can be calculated from:

$$\text{Apparent viscosity} = \text{measured stress/rate of shear} \qquad (8.3)$$

The shear stress or the apparent viscosity can be plotted against the shear rate (Figure 8.8). Using mathematical models, *e.g.* the Casson model, these plots can then be extrapolated to estimate what the stress would have been just as the chocolate started moving, *i.e.* how much force would be required to start it moving, which is the yield value. The apparent viscosity at the higher shear rates also corresponds closely with the plastic viscosity.

In order to get reproducible results, great care must be taken with sample preparation, temperature control and to ensure that the viscometer is correctly calibrated.

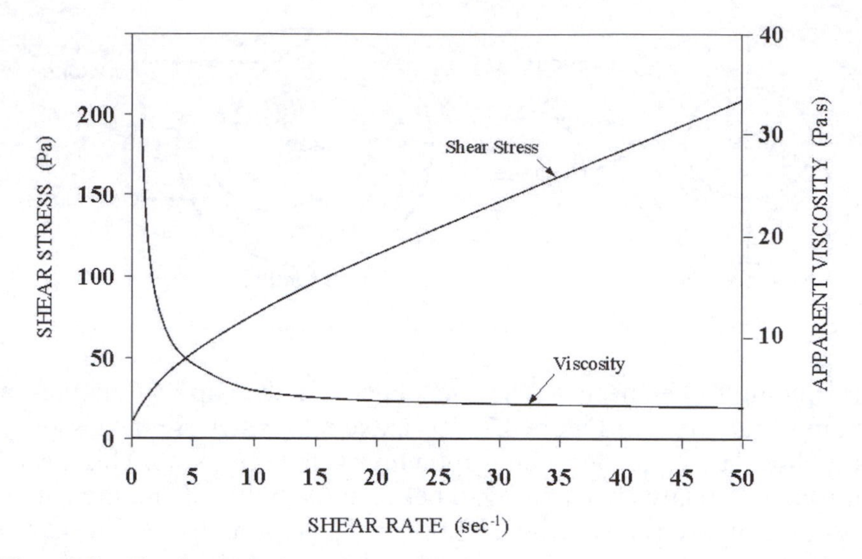

Figure 8.8 *The change in shear stress and apparent viscosity with shear rate as measured for a milk chocolate*

FLAVOUR

Flavour and texture can only be determined using trained sensory panels. There is no correct flavour, as can be seen by the wide variety of 'house' flavours, and the fact that the best selling chocolate bars are very different in different countries, *e.g.* Cadbury in the UK and Hershey in the USA. There are, however, instrumental methods that help the manufacturer to determine whether his product is correct. Very often these relate to determining off-flavours. Chocolate has a mild flavour of its own, yet it very easily picks up other often unpleasant ones. For example more and more chocolate is sold in petrol stations, yet the atmosphere in the forecourts often has a strong odour, which can be picked up by the chocolate. Likewise, great care has to be taken in ensuring the correct packaging material as chocolate will quickly take up a cardboard flavour or one from printing inks, which will make it unacceptable to the consumer.

There are many techniques which can be used to detect off-flavours and odours or to characterise different cocoas or determine their degree of roast. Instruments such as the artificial nose (this contains a series of sensors whose electrical conductivity/resistance changes according to the amount of specific molecules that are present) have been suggested for the control of cocoa liquor flavour treatment and conching. It is currently used to ensure that fats are free from taints. In this chapter liquid chromatography is described, as this technique is able to be used for a wide variety of purposes.

Chromatography is an analytical technique in which different molecules are separated using the principle that their speed of migration through a medium depends upon its size. Perhaps the best known example of this is the separation of food colours into several different coloured components by dissolving them in water and then letting the solution absorb through a filter paper or other media (see Project 13 in Chapter 10).

Many chromatographic techniques pass the sample through a medium where some of its components are either dissolved or absorbed and then record the times at which different constituents come out of the medium. If the medium is packed in a broad column, there are a lot of different paths the sample can take, so some may go straight through whereas other molecules can take much longer. This gives poor resolution, and often it is better to have very long thin columns.

In high-pressure liquid chromatography (HPLC), a solvent is continuously recirculated through tightly packed columns at pressures between 30 and 200 bar. Ultraviolet or visible photometers are often used to detect the components of the sample when they emerge. If these are fluorescent then fluorescence detectors, which are more sensitive, can be used.

The columns can be packed with solids, gels or porous or particulate materials. The pore or particle size should be kept as small as possible to increase the column efficiency. The solvent may be aqueous or of a modified hydrocarbon depending upon the sample type. The analysis can take place within a few minutes.

HPLC has been used to detect the phenols associated with 'smoky' flavour in cocoa. Ziegleder and Sandmeier[1] investigated roasted cocoa using an ultraviolet detector. They found five peaks to be of particular interest:

(1) pyrazine
(2) 2,3-dimethylpyrazine
(3) 2,5-dimethylpyrazine
(4) 2,3,5-trimethylpyrazine
(5) 2,3,5,6-tetramethylpyrazine

Peak 5 is associated with raw cocoa, whereas peak 4 correlated with the time of roasting above 120 °C. Peaks 2 and 3 increased after long roasting times at high temperatures and were an indication of over-roasting.

With the increased use of vegetable fats in chocolate (Chapter 6), HPLC has been seen as one of the possible approaches of determining their levels of use. It can also be used to detect adulteration of fats or oils.

TEXTURE MONITORING

As with flavour, sensory panels are the best way of assessing the texture of a product. Panels are, however, expensive to operate – tasters can only evaluate a few samples per day. They may also be unable to detect small changes. A manufacturer may make a change to ingredients or processing and want to know how it will affect the texture of the final product. In this case analytical techniques are especially useful, as they can evaluate a large number of samples with a high degree of reproducibility.

For chocolate, snap and hardness are two of its most important characteristics. These can be analysed by texture measuring instruments, such as the one shown in Figure 8.9. These are able to drive probes or blades *etc.* into a sample at a constant speed or force, whilst at the same time recording the resistive force or distance generated by the sample.

The relative snap of different chocolates can be measured by conducting a three-point bend test. Here bars of the test chocolates are moulded into bars of the same thickness. They are also stored for an extended period at the same temperature to ensure that this is uniform throughout the bars. Small changes in temperature can have a much bigger effect on texture than small changes in composition. In addition chocolate is a very poor conductor of heat, so it takes a long time for the centre of the bar to reach the same temperature as the surface.

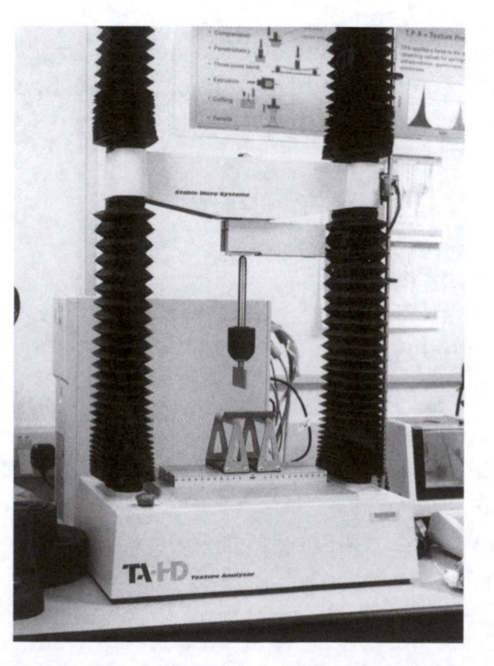

Figure 8.9 *Photograph of a TA texture measuring instrument*

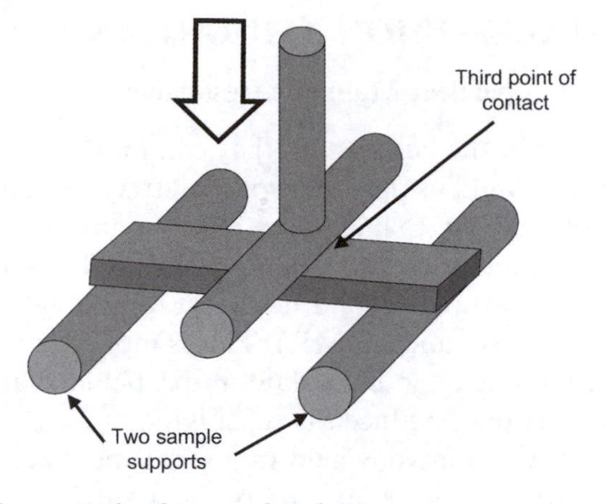

Figure 8.10 *Three-point bend test on chocolate*

The bars are then placed across two parallel supports. A probe, also parallel to these supports, is driven into the centre of the bar from above. Often the probe and support are made from rods with the same radius (Figure 8.10). The force *versus* distance graph obtained from the instrument can be related to the snap (Figure 8.11). A chocolate with a good snap will have a steep curve and a short recorded displacement as the bar will quickly break. A chocolate with a poor snap will tend to bend and so give a long slow rise.

For hardness testing a point or a ball is driven into a bar of chocolate placed on the base plate of the instrument. This can be driven in at a constant force and will give a force *versus* penetration trace, which will give a curve similar to the one for the three point bend test. A hard chocolate will once again produce a very steep curve, whereas soft chocolate will give much less resistance, so the gradient will be less steep.

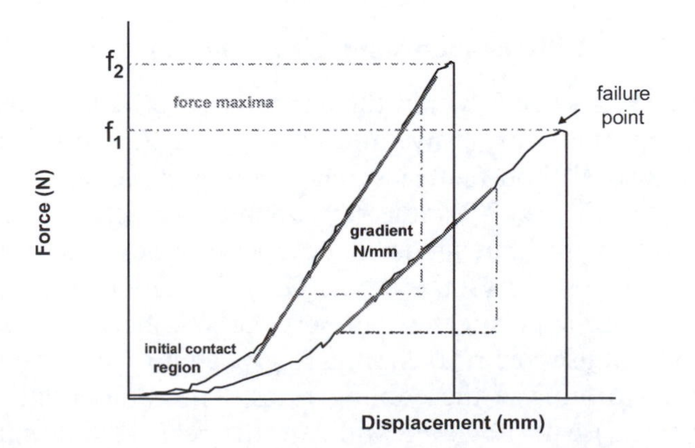

Figure 8.11 *Force versus distance curves for a three-point bend test*

CRYSTALLISATION AMOUNT AND TYPE

Nuclear Magnetic Resonance

When molecules are in the liquid state they can move much more easily than when they are bound as part of a solid matrix. If a magnetic field is applied many molecules try to line up with it. The rate at which they are able to do so once again relates to whether they are bound or free. This principle is used to measure the solid/liquid fat ratio in fat and chocolate by nuclear magnetic resonance (NMR). This is more accurate for fat, as the other solid components in chocolate, in particular naturally occurring metals such as copper, reduce the sensitivity.

In pulsed NMR, the sample is held in a magnetic field, which causes the protons in it to become polarised. A short single radio frequency pulse is then applied, which causes them to rotate about the field. After the pulse the molecules return to their original alignment over a period known as the relaxation time. This is much longer for the molecules which are bound within a solid structure. Different frequencies can be used depending upon the type of molecules being investigated, which can be either water or fat based.

This type of information can be used to build up three-dimensional pictures of what is happening inside an object in a non-destructive manner. This technique is known as magnetic resonance imaging (MRI) and is the principle of the body scanner used for medical diagnostics. This has also been used to trace the migration of soft fats through chocolate, as the soft fat is much more liquid at room temperature.[2] Figure 8.12 shows an image of a selection of confectionery products. The fat in the centres is much softer (more liquid) than that in the chocolate, so it shows up as being much lighter on the image.

Differential Scanning Calorimetry

The type of crystal present within the chocolate can be measured by X-ray analysis, but this is expensive and the equipment is not compatible with most industrial food factories. The cooling curve type of tempermeter shown in Figure 7.2 indicates whether the chocolate can be solidified satisfactorily, but tends to be operator dependent and does not indicate the types of crystal present. The operator dependence has be overcome by more sophisticated tempermeters, which operate on the same principle but use electrical cooling to give a more uniform rate, and use computers to analyse the cooling rates. The differential scanning calorimeter (DSC) can, however, indicate the relative amounts of the different crystal types that are present.

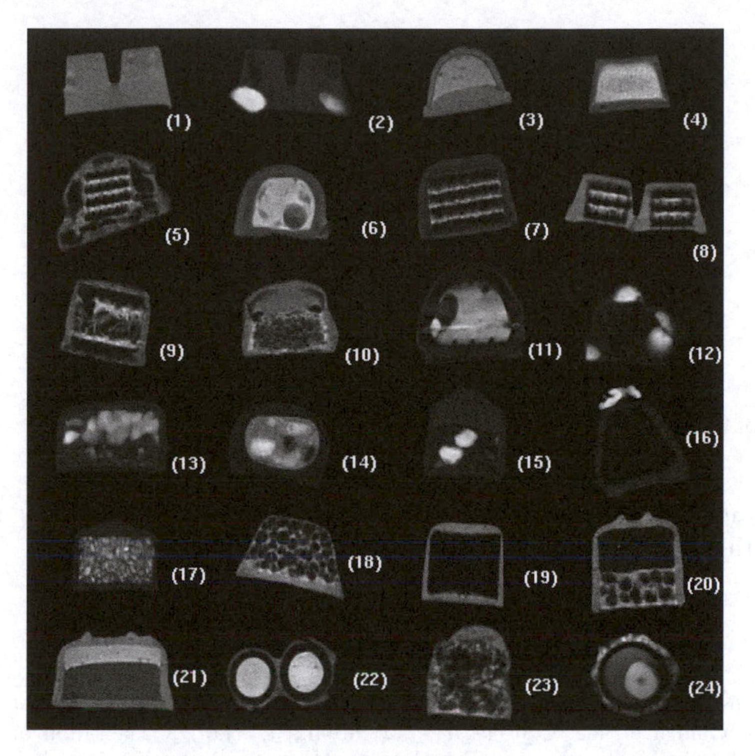

Figure 8.12 *Confectionery items as seen by MRI. The softer fat appears lighter* (Guiheneuf *et al.*[2])

This instrument is based on the principle that when a substance melts (or solidifies) a large amount of energy is required – the latent heat – but the actual temperature of the sample will be almost the same. For solid chocolate this can be used in the following way. A small (about 2–10 mg) sample of chocolate is placed in a metal container. This is then heated to give a constant temperature rise of say 5.0 °C per minute. The amount of energy required to do this is compared with that required to uniformly heat a control sample, *e.g.* a metal such as an empty container, where no melting will occur. When a particular crystalline state starts to melt, more energy is required to maintain the temperature rise, so a peak starts to rise on the energy *versus* temperature graph, as illustrated in Figure 8.13. The peak reaches a maximum at the temperature at which the rate of melting is greatest. As is also shown in Figure 8.13, several different peaks can occur for the same sample showing that more than one crystal type is present. A single peak with a maximum about 34 °C will be obtained when the chocolate is well tempered in Form V.

Sometimes it is necessary to monitor the types of crystal present during the setting stage. In this case the remaining liquid fat can be rapidly crystallised by plunging the sample into liquid nitrogen. This fat will be

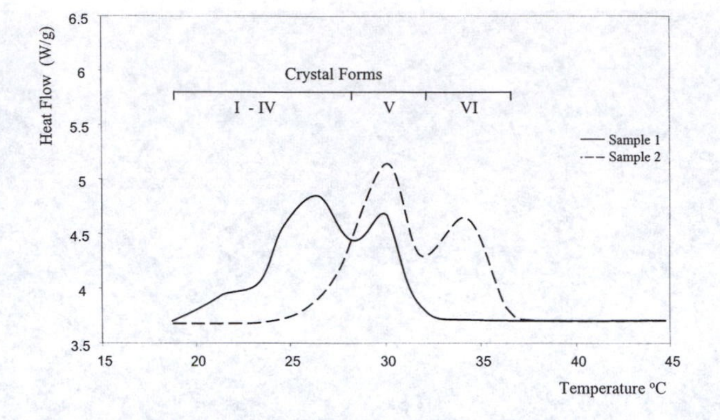

Figure 8.13 *DSC traces for different crystalline states in chocolate*

in the unstable lower melting point forms and so can be distinguished from those crystals which had already set.

REFERENCES

1. G. Ziegleder and D. Sandmeier, 'Determination of The Degree of Roasting of Cocoa By Means of HPLC', *Deutsche Lebensmittel Rundschau*, 1983, **79**(10), 343–347.
2. T.M. Guiheneuf, P.J. Couzens, H.-J. Wille and L. Hall, 'Visualisation of Liquid Triacylglycerol Migration in Chocolate by Magnetic Resonance Imaging', *J. Sci. Food Agric.*, 1997, **73**, 265–273.

Chapter 9

Chocolate Products and Packaging

Chocolate is controlled by very strict regulations in most countries. What is present is normally described on the label together with the nutritional information. Although confectionery products tend to have a relatively long shelf-life compared with most foods, care must be taken with the recipes used for the centres in order to minimise changes caused by fat and moisture migration. The packaging itself will also play a major role in determining a product's shelf-life. It is also what a consumer first sees in a shop and must therefore be attractive as well as practical.

LEGISLATION

The actual legislation that is applied varies from country to country and is changing as attempts are made to standardise it within trading areas, such as the EU. Most standards have limits as to the minimum amounts of cocoa and milk that must be present. These are referred to as milk and cocoa solids respectively. This can be a bit misleading as milk solids includes milk fat, which is a liquid at room temperature, and similarly cocoa solids includes any cocoa powder, cocoa liquor and cocoa butter. The legislation also stipulates the minimum amount of these fats that must be present and also the composition of the milk. It is not possible just to have lactose or whey instead of milk powder. In 2000 the EU agreed to legislation under which chocolate containing a higher level of milk solids (>20%) must be labelled as family milk chocolate, or its equivalent, in all countries except the UK and Eire. In addition chocolates of all types are able to contain up to 5% of a limited number of vegetable fats. These are cocoa butter equivalent fats derived from crops originating in Third World countries. The fact that vegetable fat is present must be clearly labelled.

There is also a limit to the amount of lecithin and type of other emulsifier that can be used. These may be recorded as an E number, *e.g.*

E322 is lecithin and E442 is YN, an ammonium phosphatide emulsifier. The E number shows that it has been tested and shown to be safe.

In addition there is a maximum amount of sugar that can be present and no added chocolate or milk flavours are permitted.

None of this legislation applies to products that are labelled with names such as chocolate flavoured coatings, which may contain little if any cocoa.

NUTRITION

Food is required to give us energy and chocolate is able to do this relatively rapidly. Because of this it has often been included in the food supplies for polar explorers and lifeboat rations *etc*. It also contains the three essential components of food, protein, carbohydrates and fats, together with some essential minerals. In fact 100 g of plain chocolate is able to supply 50% of the copper needed to maintain a healthy diet, whilst milk chocolate contains a relatively high level of calcium, which is widely regarded as being beneficial. Typical nutritional values for the different types of chocolate are given in the Table 9.1.

Table 9.1 *Average content for 100 g of chocolate*

	Plain	*Milk*	*White*
Energy (kcal)	530	518	553
Protein (g)	2	6	8
Carbohydrate (g)	63	56	56.5
Fat (g)	30	30	33
Calcium (mg)	63	246	306
Magnesium (mg)	131	59	31
Iron (mg)	3	2	0.2

SHELF-LIFE

The shelf-life of a chocolate confectionery product is determined by when its taste, texture or appearance changes by an amount to make it unattractive to the consumer. In many cases this is when it turns white due to fat bloom. This change in appearance is often coupled with a hardening in texture and a slower melting, caused by changes in the fat crystallisation (usually the Form V to Form VI transition). These changes are slowed down by using good storage conditions and the use of special fats and emulsifiers in the chocolate or soft fat filling, as described in Chapter 6.

The surface can also look unattractive if it is badly cracked. This was noted to occur for panned products, when the centre expands more than the chocolate coating due to changes in temperature. This will also, of course happen for moulded and enrobed products, and so a constant temperature is desirable. However, moisture transfer can also cause the centre to expand, especially for some baked centres.

Although a continuous chocolate coating slows down moisture transfer, water molecules will migrate slowly through it, so expansion of this type of centre will eventually occur. Once a crack has appeared, the moisture transfer increases rapidly and the product's appearance will deteriorate rapidly, as will the eating properties of the centre.

Wafer-like centres take up moisture and become soggy, losing their crispness and becoming unpalatable. With high-moisture centres such as caramel or fondants, drying out can occur making them hard or gritty, so once again moisture transfer is a problem. This can be made worse when both types of material are present in the centre and they don't have a chocolate or fat layer between then to slow down the transfer. An example of this is Lion Bar with a wafer centre and a caramel coating.

Figure 9.1 illustrates some of the fat and moisture migrations which take place and which lead to the deterioration of the product quality. Even water going into the chocolate can change the texture and make it appear staler, although this is relatively unimportant compared with the wafer/caramel reaction. The driving force that governs the rate of these changes is the difference in equilibrium relative humidity (ERH) (= 100 × the water activity). This can be measured for each of the components by placing them individually in sealed containers with a small air space above them. This air will rapidly take up, or give out, moisture to the ingredient so that the two rapidly reach an equilibrium. This is then measured by a probe calibrated to read relative humidity.

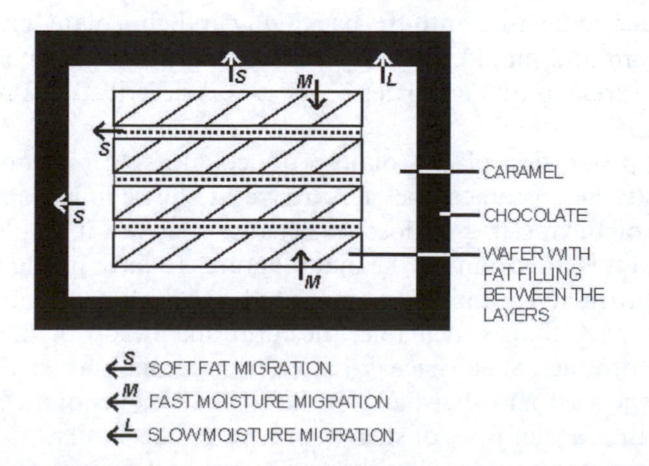

Figure 9.1 *Illustration of some of the causes of deterioration in a product containing wafer, caramel and chocolate*

The ERH will depend very strongly upon the moisture level within the ingredient, its composition and also the temperature. For the product the important temperature is that in which it is going to be stored, which might be 15 °C for the warehouse and 22 °C for the shop.

In order to minimise moisture transfer the ERHs of each of the three components must be as close together as possible. The wafer can only have a limited amount of water present before the texture becomes too soft, and the caramel will be too hard if too little is present. This means that other components must be modified to change the ERH. Small amounts of ingredients which are highly hydrophilic (water loving) can for instance greatly increase the ERH.

The form of the packaging is also important as, unless a good moisture barrier material is used, the humidity of the storage area can play a big role, with dry centres becoming soggy under high humidity and moist ones becoming hard or gritty in dry conditions.

PACKAGING

It is very important for the packaging to be chosen to take into account the product and the conditions under which it is going to be stored. At first sight it might seem best to package a high-moisture centre with a water-impermeable barrier. This is not always the case, however. Some years ago a Turkish Delight product was wrapped using such a moisture barrier only for a lot of the product to go mouldy. The reason behind it turned out to be connected with the fact that the temperature varied during storage. Many shops are warmer during the day than at night. The moisture from the Turkish Delight eventually migrated through the chocolate so that the air within the wrapper was saturated during the day, when the room was hot. When the temperature fell below the dew point, water condensed onto the packaging and chocolate, giving rise to sugar bloom and mould. A method of testing packaging materials for their moisture and odour barrier properties is described in Project 14 in Chapter 10.

A high proportion of chocolate products are sold on impulse. That means that the customer does not arrive at the selling place with the intention of buying it, but does so because they see it displayed. This means that it is important to the manufacturer to have products that are packaged attractively and show up due to bright colours *etc*. In addition, the more space that is available, the more likelihood of the customer seeing the product. Shelf space is limited in supermarkets, so alternatives are used such as bags that hang on hooks or large containers that are placed at the ends of rows of shelves or near to checkouts.

There are many types of packaging, some just providing protection against dirt whilst others provide a good barrier against external odours.

Figure 9.2 *Distinctive brand packaging as illustrated by Smarties and Toblerone*

Others become associated with the product, *e.g.* the Smartie tube and Toblerone (Figure 9.2). In order to meet production requirements (over 17 000 Smarties are eaten in the UK every minute), these tubes are filled and sold by weight. As not all Smarties are the same size this can give some interesting mathematical problems, when special new Smartie types are introduced (see Project 12 in Chapter 10). In addition there is pressure to reduce the amount of packaging material used. This is partly for environmental reasons and also because it increases transport costs, as well, of course, so as to reduce the cost of the packaging material itself. Reducing the thickness of the material is one way of doing this. The size of the wrapper is also important. Because so many chocolates are sold, only a few millimetres less of packaging per sample can save a lot of wrapping material.

Packaging machines are also very complex and expensive. In the 1950s large amounts of labour were required to package products, but now individual machines may wrap several hundred bars per minute. These are expensive, however, and it is important that they are kept running as long as possible. This not only means a continuous supply of product of the correct size into the machine (too large pieces may block it), but also that stops do not occur when the wrapping material roll runs out (most modern machines go automatically on to a new reel) or if the quality of the wrapping material is not consistent and it stretches or breaks.

Foil and Paper Wrap

The traditional packaging for moulded chocolate blocks and tablets is foil and paper (Figure 9.3). The aluminium foil provides some protection against dirt, insect infestation and taint, whilst the paper can be brightly

Figure 9.3 *Traditional foil- and paper-wrapped chocolate products*

coloured with the product name and also has the required legal and nutritional information printed on it. The thickness and size of the foil and paper can be minimised, and both these materials are easy to recycle.

In order to transport them, generally two to six dozen bars are placed in a cardboard box, known as an outer. These outers are made of paperboard and can be decorated on the outside so that the retailer can sell directly from them (Figure 9.4). Where extra shelf-life is required, particularly in hotter more humid climates, this outer can have a barrier film put around the outside. Outers are combined together in shipping

Figure 9.4 *Confectionery products displayed in an outer*

cases and put on pallets. In some cases outers are palletised directly. The pallet load is secured by having plastic film stretched tightly around it.

Foil is also used to pack seasonal novelties such as Easter Eggs and Father Christmases. For the latter the packaging is more complicated as the printed design on the foil must fit with the markings on the chocolate.

Flow-Wrap

A large proportion of chocolate confectionery is sold as a countline. This is an individual unit often purchased and eaten by the consumer as a snack and not part of a meal. The majority of these are packaged using flow-wrap. This has the advantage that large numbers of items can be wrapped by a single machine, over 500 a minute in some cases. The pack is sealed tightly and with the appropriate choice of material can be a very good barrier to moisture and odours. A similar procedure (although not really fitting the flow-wrap definition) is also used to make larger bags which contain a selection of smaller sweets, which themselves may or may not be wrapped. Both types of flexible packaging are shown in Figure 9.5.

Wrapping material is supplied as long rolls of printed material. Great care must be taken with the type and application of the inks, otherwise they can impart an unpleasant flavour to the chocolate. Space is often left so that the 'best before' date can be added on each individual item as it is packaged. This may be done by ink jet coding or laser burning. A wide range of materials can be used including thin coextruded or white cavitated polypropylene and film/foil/ionomer laminates

The flow-wrap is initially unwound from the roll and made into a tube (see Figure 9.6) by sealing it with heat or pressure. The product is fed into the tube, which is then cut into the required lengths. The open ends are

Figure 9.5 *Flow-wrapped confectionery products*

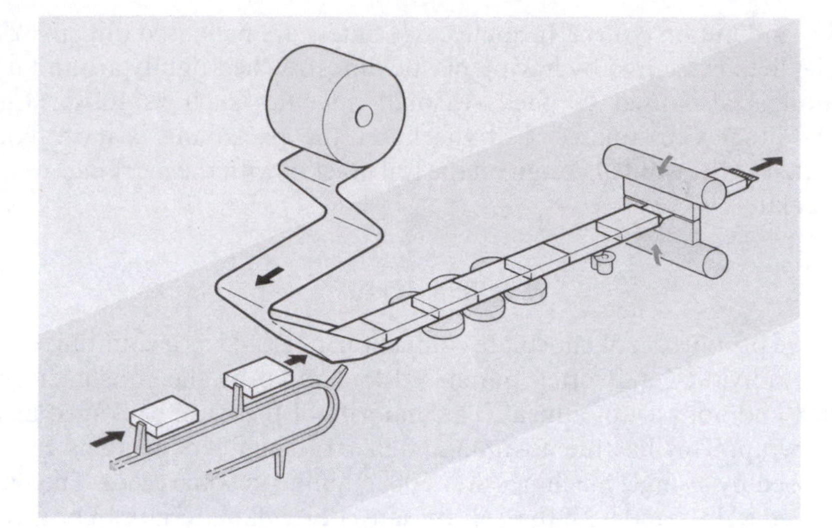

Figure 9.6 *Schematic diagram of the flow-wrapping process*

then sealed, once more by heat or pressure. Sealing by pressure requires cold adhesive to be applied to the reverse side of the wrapping material during the printing process.

This sealing is particularly important where it is necessary to have a good moisture or taint barrier. Where heat is used to do this, sufficient time must be allowed for the sealing layers to fuse together, which not only restricts the packaging rate but also introduces the possibility that the product itself will be damaged by the heat. A method of cold sealing has therefore become increasingly popular. This is based on natural rubber latex combined with resins. Not all packaging films will accept this cold seal, however, and a release lacquer or film may be required to avoid blocking on the reel. Fast packaging with very good barrier properties has been obtained with cold seal and metalised film (film with a microscopical aluminium layer deposited under vacuum).

Robotic Packing

Placing individual sweets inside a selection box is very labour intensive (Figure 9.7). Although this is still done by hand in many factories, in some it is carried out by robots. Image recognition systems identify that the sweets are the correct shape and which way they are pointing. An arm, often with a suction system at the end, picks up the sweet and places it in the proper location in the box. Early robotic systems were very restricted in that they were difficult to reprogram when sweets were changed within the selection. Modern, more powerful computing systems have overcome this problem.

Figure 9.7 *Robots packing boxes of chocolates*

Much simpler robots are used to manipulate boxes or packages of sweets and put them into outer packages. This type of machinery has dramatically reduced the number of people working in the chocolate industry and, together with the new processes described earlier, has transformed it from a high manpower craft industry to a higher production, highly scientific and technical one.

Chapter 10

Experiments With Chocolate And Chocolate Products

The aim of this chapter is to outline a series of projects which can be carried out to demonstrate chemical, physical or mathematical principles. They have been designed to only use relatively simple apparatus and should be able to be adapted for students over a wide range of ages.

PROJECT 1
AMORPHOUS AND CRYSTALLINE SUGAR

Apparatus:
Beaker.
Magnetic stirrer.
Thermometer sensitive to better than 0.1 °C.
Balance capable of reading to at least 1 g.
Granulated sugar.
Skim milk powder.
Boiled sweets, *e.g.* Foxes Glacier Mints (*NB* not pressed or tabletted sweets such as Polos which are more crystalline).

Aim:
To show how amorphous and crystalline sugar differ when they dissolve in water. The crystalline sugar causes the water to cool down as energy is required to separate the molecules (heat of solution). Amorphous sugar is in an unstable state, however, and gives out energy when it changes to its stable lower energy crystalline state. This means that there is spare energy so the water becomes warmer.

Procedure:
Pour 10 ml of water into a beaker and place on the magnetic stirrer.
Place the thermometer in the water and let it continue stirring until the temperature is constant.

Break the boiled sweets into small pieces. (This can be done by placing it within a material bag and crushing it with a hammer. **Care** – take appropriate precautions.)
Weigh out about 10 g of granulated sugar and of the crushed amorphous material.
Drop this quantity of granulated sugar into the water and record the temperature for the next five minutes.
Repeat the procedure using the amorphous sugar. This time the temperature should rise.

The experiment can also be tried with skim milk powder, *e.g.* Marvel. This contains lactose, which is normally in an amorphous state owing to the rapid spray drying process. This normally gives a much greater temperature rise than the boiled sweet.

PROJECT 2
PARTICLE SEPARATION

Apparatus:
Several sheets of paper.
Scissors.
A small narrow glass jar with a lid.
A stop watch.
Dried peas, rice, lentils and sunflower seeds.

Aim:
To investigate some of the principles that are used to separate cocoa beans from stones and cocoa nib from shell and that are the cause of the segregation of some centres during chocolate panning.

Procedure:
Separation by Vibration.
Partly fill the narrow jar with dried peas.
Pour some of the rice so that it forms a layer on top.
Stir so that the two are well mixed and put the top on the jar.
Shake gently with a slightly swirling action.
The peas should then come to the top and most of the rice at the bottom. as shown in Figure 10.1.
Turn the jar over and shake again.
The peas should once more be at the top.
Repeat with other materials such as lentils and sunflower seeds.
NB Shaking the jar on its side can sometimes be more effective.

Figure 10.1 *Picture of peas and rice mixture in a small jar after shaking*

Separation by Falling Speed

Cut one of the pieces of paper in half. Then one half of this into strips about 1 cm in width.

Screw another piece of paper into a ball.

Hold the full sheet of paper horizontally at a height of at least 2 m. Time it as it falls to the ground. Repeat this several times and determine the average falling time.

Repeat this with the screwed up ball.

(This will fall much faster, and shows why the plate-like shell of cocoa beans can be separated by falling speed – or upward suction – from the spherical nib.)

Repeat the experiment with the long paper strips. These will fall faster than the complete sheet, but slower than the ball. The falling speed of a fibre is governed by its diameter and not its length. So providing that they are long enough to be fibres (the length should be at least 10 × the diameter) all the thin strips of paper, with the same width, should fall at about the same rate.

PROJECT 3
FAT MIGRATION

Apparatus:

Two desiccators or sealed containers.

Printing/photocopier paper.

Silica gel.

Cocoa butter as obtained from Project 4. (Other hard fats should give the same results.)

Confectionery wrapping material.

Aim:

To show that moisture affects the release of cocoa butter from the cells within the cocoa nib.

Procedure:

Put silica gel in the base of one desiccator. Leave some paper in a relatively moist environment.

Put some paper into the desiccator and leave for several hours.

Remove the papers and put a drop of fat on each.

The fat will remain as a globule on the wet papers but spread out and sink into the dry ones.

This can be repeated with the PVC type of film used to wrap a lot of confectionery. This will not let the fat sink into it under either conditions and is why it has replaced cellulose as a packaging material for a lot of confectionery.

PROJECT 4
COCOA BUTTER SEPARATION

Apparatus:

2 × 1 litre beakers.

Oven.

Knife

Bars of chocolate.

Stirrer.

Aim:

To separate cocoa butter from the chocolate and to show the effect of lecithin as an emulsifier in a fat in water system.

Procedure:

Use the knife to scrape the chocolate into small flakes.

Three-quarters fill one beaker with water at around 60 °C. **Care: this is hot.**

Slowly pour the flakes into the water and let them settle on the bottom.

Gently agitate the flakes using the stirrer. **Do not mix vigorously.**

Place the beaker in an oven at 50–60 °C for 12 hours. **Care: it is hot.**

Remove from oven and let it come to room temperature.

This should form a yellow fat layer on top, which is a mixture of cocoa butter and milk fat.

Repeat the above but stir very vigorously when the flakes are in the beaker. **Care: it is hot.**

The lecithin in chocolate is attached to the sugar. When the sugar dissolves it can go into the water. The fat is the continuous phase in chocolate and if it remains in large globules when it melts then the layer of fat will form. If it is stirred vigorously the fat forms small droplets which can be coated with lecithin. These remain suspended within the water as an emulsion.

PROJECT 5
CHOCOLATE VISCOSITY

Apparatus:
Temperature controlled cabinet (40 °C).
Thermometer accurate to 0.5 °C.
Tall beaker.
Ball fall viscometer (a design for its construction is given in Figure 10.2)
Stopwatch.
Plastic funnel.
Balance capable of reading to 1 g.
Bars of chocolate.
Sunflower oil.

Aim:
To show the effect on the flow properties of chocolate of fat and moisture additions.

Procedure:
Effect of Fat
Melt the chocolate and keep it, the sunflower oil and the viscometer, plus any glassware, in the cabinet at 40 °C and allow it all to come to temperature. Check using the thermometer.
Fill the tall beaker almost to the top.
Use the stop watch to record the time it takes for the ball fall viscometer to sink between the two markings on the bar or wire.
Repeat this four or five times, disregarding the first reading.
Repeat with different sizes of ball, if available.

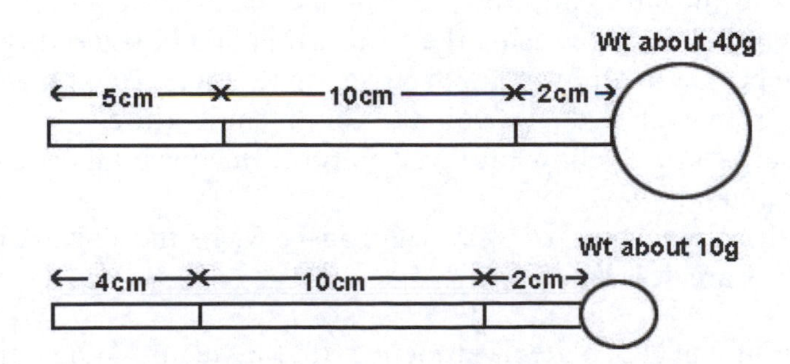

Figure 10.2 *Schematic diagram of ball fall viscometer*

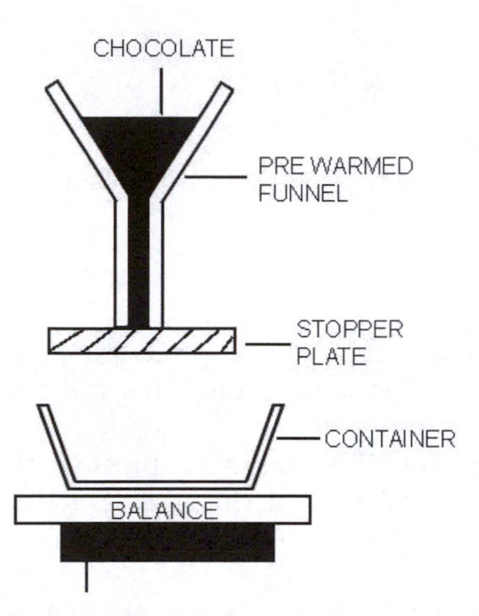

Figure 10.3 *Set up for flow cup viscometer*

Place the funnel on a stand so that any liquid in it will flow into a container on the balance (Figure 10.3).
With a finger or card under the outlet, partly fill the funnel.
Measure the time it takes for about half of the chocolate to flow out.
Refill and repeat the procedure several times.

Add sunflower oil to the chocolate so that it makes up about 3% of the mixture, stir very thoroughly so that the mixture is uniform (a food processor can be used if available) and repeat both types of measurement.
This can then be repeated with other levels of addition.
What effect does the extra fat have? Does it affect the ball fall viscometer (predominantly yield value) as much as the flow cup (mainly plastic viscosity)?

Effect of Moisture
Repeat the procedure as for the sunflower oil, but this time use water.
How do these results compare with the previous ones? See also Project 8.

PROJECT 6
PARTICLE SIZE OF CHOCOLATE

Apparatus:
Microscope (if possible with polaroids).
Sieve (about 50 microns mesh).
Balance capable of reading to 0.1 g.
Micrometer capable of measuring 40 microns.

Sunflower oil.
Several different brands of chocolate.

Aim:
To investigate the largest particles within chocolate and see whether particle size affects its texture/flavour.

Procedure:
Microscopy
Place a small amount of chocolate on a microscope slide and disperse the solid particles in a transparent liquid that will not dissolve them, *e.g.* sunflower oil.
Put a coverglass on top of the same and press gently to remove the air. (Pressing too hard will move all the particles to the edge.)
Calibrate the microscope to be able to determine the size of particles greater than 20 microns.
View the sample and determine the size of the biggest particles.
If the microscope is a polaroid one, cross the polaroids so that the field of view is dark when no sample is present. Re-examine the sample. Crystalline sugar is birefringent, which means that it will show up as a bright image (Figure 2.10), whereas the cocoa and milk particles will remain dark. It is therefore possible to see whether the biggest particles are sugar or not.
Repeat for all the different chocolates.

Sieving
Disperse about 10 g of liquid chocolate in about 100 ml of warm sunflower oil.
(If an ultrasonic bath is available, this will help to disperse the particles.)
Pour the suspension through the sieve.
This will retain the biggest particles.
These can be defatted and weighed, or viewed on the microscope as described above.

Micrometer
Prepare a concentrated suspension of chocolate in sunflower oil.
Place a drop of the suspension on one of the jaws of the micrometer.
Screw the jaws of the micrometer together and take a reading.
This should be done several times and the reproducibility of the technique determined, as it is very easy to screw too hard and break the particles.

Flavour
Get at least five people to evaluate the samples. The samples should each be given a value between 1 and 10 for, grittiness, creaminess and cocoa flavour.

Determine whether there is a correlation between any of the size measurements and these three sensory parameters.

PROJECT 7
EFFECT OF LECITHIN

Apparatus:
Food processor or mixer.
Icing sugar.
Sunflower oil.
Lecithin (obtainable from many health food shops).

Aim:
To show the large effect on viscosity of lecithin in a sugar fat mixture.

Procedure:
Mix five parts of sunflower oil with two parts of icing sugar in a food processor for five minutes.
Check the viscosity by estimating how hard it is to stir, or use the ball fall viscometer (Project 5) if it is thin enough.
Add 5% of lecithin (warm slightly if it is very thick) and mix in the processor for a further 2 minutes.
Remeasure the viscosity.

PROJECT 8
CHANGING THE CONTINUOUS PHASE

Apparatus:
Oven set at 40 °C.
Stirrer (ideally connected to a torque or a power meter, but stirring by hand and estimating how hard it is to mix is possible.)
Bars of chocolate.

Aim.
To show that small additions of water to chocolate make it thick by ‘sticking’ the sugar together (Project 5), but larger additions make it thinner again as the water is able to form a continuous phase.

Procedure:
Melt the chocolate for at least 3 hours in an oven at 40 °C.
Stir and estimate or record how hard it is to do so.
Add 2% of water, mix thoroughly and then determine how hard it is to stir it.
Repeat the procedure until 30% of water is present.
Plot the ‘stirability’ against the water content.

PROJECT 9
CHOCOLATE TEMPER

Apparatus:

Thermometer including the range 20–32 °C and accurate to better than 0.5 °C. An electrical thermometer is preferred as it will have a much smaller heat capacity than a glass one.

Less accurate thermometer with a range including 50 °C

Two beakers and stirrers.

Test tube with stopper and stirrer.

Stop watch.

Hot plate.

Bars of chocolate.

Aim:

To show that pre-crystallised chocolate sets differently to untempered chocolate.

Procedure:

Melt about 30 g of chocolate and heat in a beaker to 50 °C. **Care: this is hot**. Stir occasionally and maintain this temperature for about 30 minutes, to ensure that no fat crystals remain.

Pour about 10 g into the bottom of a test tube and put the stopper containing the thermometer and stirrer into the top, as shown in Figure 10.4.

Put the test tube in a beaker of cold water and plot the change in temperature with time once the chocolate is below 32 °C. Stir intermittently as the chocolate sets.

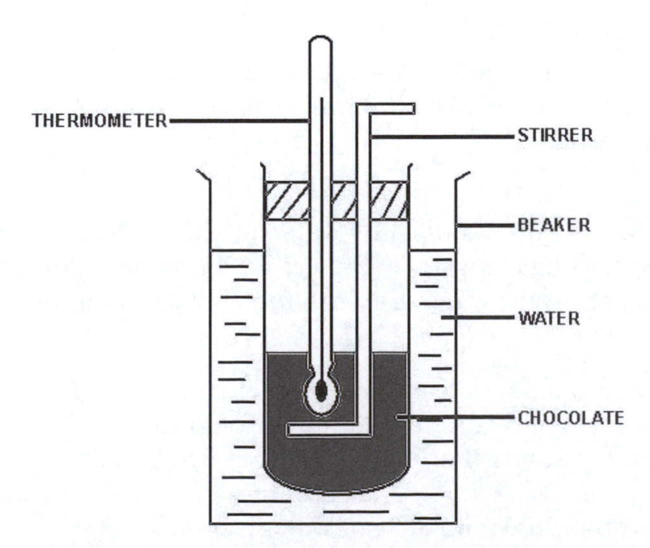

Figure 10.4 *Apparatus to monitor the setting properties of chocolate*

Cut or grate about 5 g of chocolate into a powder.
Repeat the above procedure until the chocolate is about 35 °C.
Add about 3 g of the powdered chocolate and stir vigorously.
Start measuring the rate of cooling.
(The powdered chocolate should 'seed' the chocolate with Form V crystals, which should then make it set and the temperature rise much more rapidly than was the case with the first sample.)
The experiment can then be repeated using iced water or with different amounts of chocolate.

PROJECT 10
HARDNESS MEASUREMENT

Apparatus:

Temperature controlled cabinet.
Set of half kilogram and kilogram weights.
Retort stand.
Counter sink, *i.e.* a broad metal rod with a conical tip at one end.
Travelling microscope (although it should be possible to do the experiment with a ruler and a hand magnifier).
Bars of chocolate.

Aim:

To demonstrate that small changes in temperature can have a big effect on the hardness of chocolate.

Procedure:

Store bars of chocolate at different temperatures for at least 12 hours. These could include in a refrigerator, warm room, and controlled cabinets at 24 °C and 28 °C.
Set up the retort stand so that it will hold the rod (through a loose-fitting tube) vertically above the sample being tested (Figure 10.5).
Take one of the samples and place it gently below the point. The samples should be removed from the storage conditions just before testing, so that the fat doesn't melt or harden.
Carefully place a weight on the top of the rod and balance it there for a few seconds.
Remove the weight and lift the point of the rod out of the chocolate.
Move the sample slightly to one side and put the point on the surface.
Carefully place a bigger weight on top of the rod.
This procedure should be repeated several times, a note being made as to which mark corresponds to which weight.
Measure the diameter of each of the marks using a travelling microscope.
Relate the diameter readings to the weight applied and also to the temperature of storage.

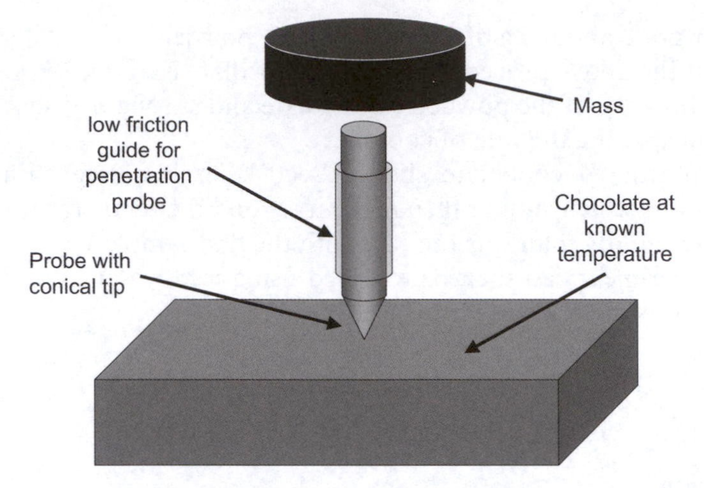

Figure 10.5 *Apparatus for determining the relative hardness of chocolate*

PROJECT 11
CHOCOLATE COMPOSITION AND PRODUCT WEIGHT CONTROL

Apparatus:

Balance capable of weighing to at least 0.1 g.

Bars of solid chocolate and countlines.

Aim:

To look at the chocolate composition of products on the market.

To investigate the weight of the bars compared with their declared weight and to see if there is any difference between enrobed and moulded products.

Procedure:

Survey the chocolate bars being sold in the shops with respect to their cocoa and milk solids. Also note any that indicate the country of origin of the cocoa. These are most likely to be found on plain chocolate, especially organic products.

Buy at least ten samples of individual products, as far as possible from different shops. (Samples from the same shop are likely to have been made at the same time, so differences will be less). The samples should include moulded chocolates like solid bars and countlines such as Mars Bar, Drifter, Crunchie and Lion Bar.

Weigh all the samples and plot a graph of the number of bars within set weight ranges. Calculate the mean weight, the standard deviation and coefficient of variation (standard deviation/mean) for each product and compare this with the declared weight.

The coefficient of variation of the moulded solid bars would be expected to be smaller than that for enrobed ones with centres, as not only is enrobing normally less precise, but there is also often a weight variation between the centres.

PROJECT 12
DISTRIBUTIONS AND PROBABILITIES

Apparatus:
At least 20 tubes of Smarties or a similar coloured product.

Aim:
To illustrate the importance of number distributions and probabilities.

Procedure:
For each of the tubes record the total number of sweets as well as the number of each individual colour. Smarties are made as individual colours which are then mixed and the tubes filled by weight not by number. This means that certain colours will be missing and because not all the Smarties are the same weight, different tubes will often contain a different number of sweets.

If a new colour is introduced, what proportion must be used in the original mixture to ensure that there is at least one of the new type in each tube? What is the probability of a tube containing only one colour?

Plot the number distribution (histogram) for the contents, using the fact that there is a range of values for the total number of Smarties within the tubes. Plot a pie chart showing the percentage of each colour of Smartie present (Figure 10.6).

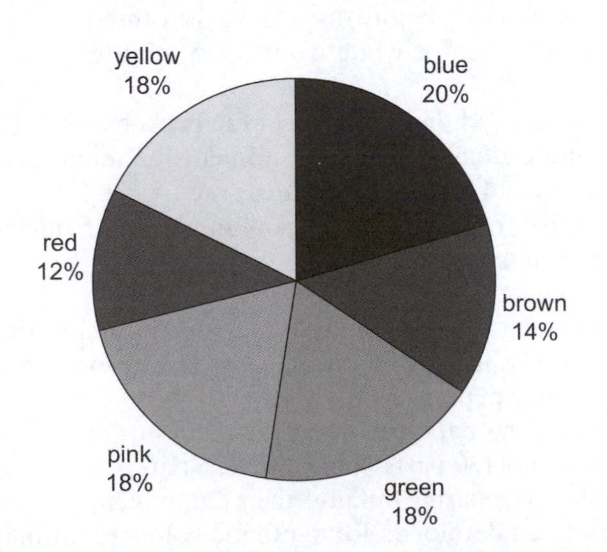

Figure 10.6 *Pie chart of the number of Smarties with different colours in 48 different tubes*

If, as a promotion, the Marketing Department offer 10% free, how many more Smarties on average will you get? Is it possible to have 10% extra weight and yet have fewer Smarties?

PROJECT 13
CHROMATOGRAPHY OF COLOURS

(Thanks to Nestlé Rowntree Consumer Services Department)

Apparatus:

Eye and hand protection are advised, as chemicals are involved.

Coloured sweets such as Smarties.
100 ml beaker.
Water bath.
Hot plate.
1% ammonia solution.
1 m of white wool.
Ethanoic (acetic) acid.
Ethyl acetate.
2-propanol.

Aim:

To separate a food colorant into its individual components.

Procedure:

Take at least seven sweets of the same pale shade or four for the darker ones.

Place in 10–15 ml of hot water in a 100 ml beaker and allow the surface and colour to dissolve, before discarding the centres.

Boil approximately 1 m of white wool for 5 minutes in 1% ammonia solution.

Rinse in cold water and then immerse in the beaker with the dye extract.

Acidify with dilute ethanoic acid and simmer for 5 minutes.

Remove the wool and rinse in cold water.

Re-extract the dye from the wool by soaking in 10–15 ml 1% ammonia solution for 5 minutes.

Remove the wool.

Evaporate the ammonia solution to dryness on a water bath.

Add two drops of water to the extracted dyes and spot on to a strip of chromatographic paper.

Place one end of the paper in a solvent made up of ethyl acetate (40 parts), 2-propanol (30 parts) and water (25 parts).

The colours should separate out into their components.

This technique was developed for synthetic colours and may not work well for riboflavin and cochineal.

PROJECT 14
THE EFFECTIVENESS OF DIFFERENT PACKAGING MATERIALS

Apparatus:

Balance capable of reading to 0.1 g.
Desiccator.
Silica gel.
Aluminium or glass dishes about 5 cm in diameter and 1 cm deep.
Wrapping material from confectionery including, aluminium foil, paper and plastics.
Sealing wax or odour free adhesives.
White chocolate buttons.
Peppermint oil.

Aim:

To assess the relative effectiveness of different packaging materials as barriers to moisture and odour transfer.

Procedure:

Moisture Barrier

Fill the base of several dishes with the same weight of silica gel.
Cut a circle of the packaging material, slightly bigger than the dish and seal it over the top using the wax or an adhesive. Be careful that there is a good seal all the way around. Then record the weight of each of the dishes.
Put water in the bottom of the desiccator and then place the dishes inside it as in Figure 10.7.

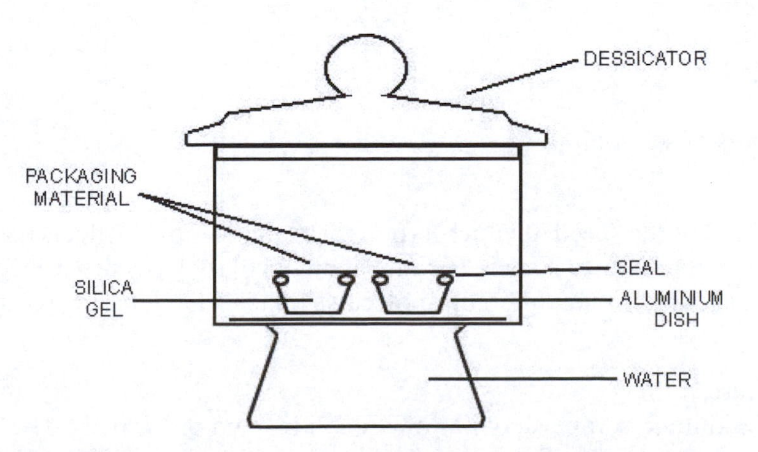

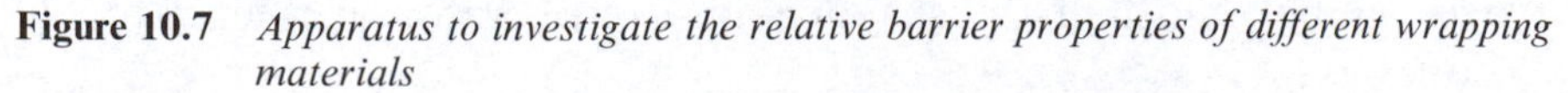

Figure 10.7 *Apparatus to investigate the relative barrier properties of different wrapping materials*

Take the dishes out every day and reweigh them.
The poor moisture barriers will pick up weight relatively quickly. Eventually the weight will become constant again because the silica gel won't be able to pick up any more moisture.

Odour barrier

White chocolate is used as it has a relatively bland taste.
Place about six buttons in each dish, but this time have at least five dishes for each type of packaging material.
Once again seal the packaging material on top. This time there is no need to weigh them.
Put the peppermint oil in the base of the desiccator and then the dishes on the tray.
At intervals (*e.g.* 1, 5 and 14 days) take out one dish for each type of packaging material.
Get five different people to taste the buttons and to score them for the intensity of peppermint flavour on a scale of 1 to 10.
Plot this intensity *versus* time for the different packaging materials.

Leak test

Place a flow-wrapped sample or bag under the surface of water in a container and squeeze gently.
Count the rate at which bubbles come to the surface.
If the rate is such that you can count aloud the number of bubbles, then the packaging is satisfactory even though some leaks still exist

PROJECT 15
VISCOSITY AND FLAVOUR

Apparatus:

Bars of chocolate.
Refrigerator
Knife.
Two pots of set yoghurt

Aim:

To show that the speed at which a food melts in the mouth affects its taste as well as its texture. This is because the chocolate viscosity affects the speeds at which the different molecules reach the flavour receptors (Figure 5.1).

Procedure:

Use the knife to scrape shavings of chocolate from one bar.
Place one bar and half the shavings in a refrigerator or freezer for 24 hours.

Keep the other half of the shavings with another bar of chocolate in the room.

Taste all four samples recording their hardness, speed of melt, creaminess and cocoa intensity. Although all were originally the same, large differences should be recorded.

A similar effect can be obtained from set yoghurt.

Stir one pot vigorously so that it becomes a thin liquid, then compare the taste of this with the other thick sample.

PROJECT 16
HEAT RESISTANCE TESTING

Apparatus:

Part a

Oven set at 32 °C.

Filter paper.

Several different brands of chocolate.

Knife.

Part b

Refrigerator.

Balance (accurate to three places of decimals).

Aim:

Chocolate melting and sticking to the packaging is a major problem in warm or hot climates. This experiment shows one method of measuring how badly this is likely to happen and also that the chocolate type affects its heat resistance.

Procedure:

Part a

Mark the filter paper so that it is divided into small squares.

Cut the chocolate into equal sized squares.

Place the samples in the middle of the filter paper in the oven at 32 °C for 2 hours.

Remove from oven and take the chocolate off the filter paper.

Count the squares that are fat stained.

The more squares that are effected the easier the chocolate melts. It may be possible to relate this to the smoothness of the chocolate. Why is this?

Part b

Break off four pieces from the bar.

Determine the weight of the four pieces.

Put two pieces one way up on the filter paper, the other two the other way.

Place these in an oven at 32 °C for 2 hours.

Remove and put immediately in a refrigerator for at least one hour.
Remove the chocolate and reweigh it.
Determine the fat loss.
The effect of oven temperature and chocolate size and shape can also be investigated.

PROJECT 17
COEFFICIENT OF EXPANSION

Apparatus:
Water bath.
Flask fitted with stopper and with a glass tube through the middle (Figure 10.8).
Thermometer accurate to at least 0.5 °C.
Chocolate.
Sugar solution.

Aim:
Some chocolate confectionery products crack because two of the components expand at different rates when the temperature changes.

Procedure:
Melt the chocolate at about 40 °C for several hours, then cool to 38 °C.
Set the water bath to 38 °C.
Pour the chocolate into the flask so that it is almost full and push the bung and tube into the top. **Care: wear gloves in case the tube breaks.**

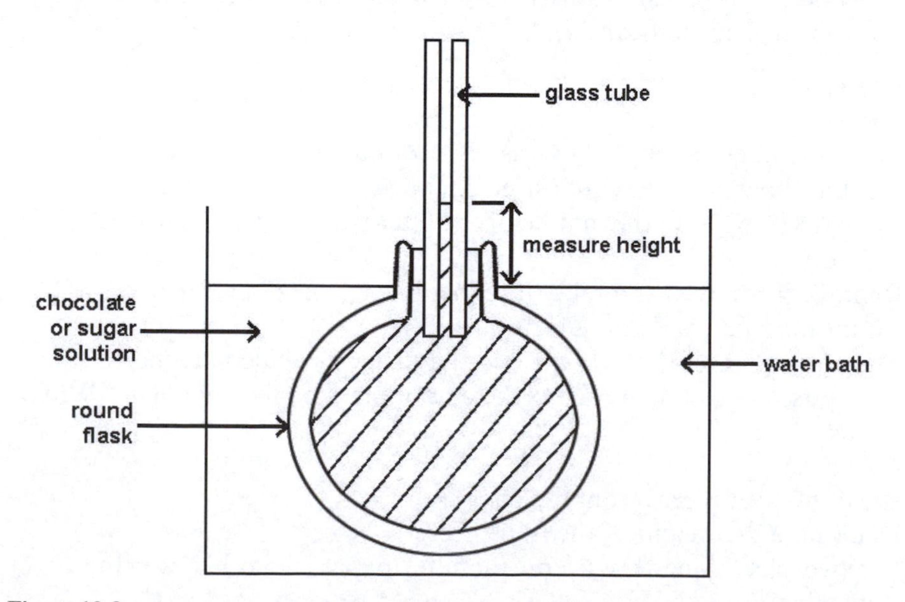

Figure 10.8 *Apparatus to compare the relative coefficients of thermal expansion of chocolate and sucrose solutions*

Place the flask in the water bath so that the neck is just above the water.
Allow the system to come to equilibrium and then measure the height of the chocolate up the tube from the bung.
Raise the temperature by 2 °C and leave for 20 minutes, then remeasure the height.
Repeat up to 50 °C.
It will then be possible to plot the height of the chocolate column against temperature.
Repeat the experiment but using sugar solution.
The two curves should be different.

PROJECT 18
THE MAILLARD REACTION

Apparatus:
Glucose.
Valine (can be purchased from national chemical suppliers).
Small beaker (100 ml).
Hot plate.
Sunflower oil.

Aim:
To try to develop some of the chocolate aroma that is formed during roasting.

Procedure:
Dissolve about 3.6 g of glucose and 0.6 g of valine in 20 ml of water.
Add about 2 ml of sunflower oil.
Heat to near boiling point for about 15 minutes (try to simmer and stir occasionally). Smell the aroma. **Care: very hot – use gloves and safety glasses, and take extra caution if it is spitting.**
Test for different combinations of temperatures times and relative concentrations.
Valine is only one of the many amino acids that are present in cocoa, so it will not be possible to generate the full aroma of chocolate.

Subject Index